A30125001128049B

KV-577-636

Books are to be returned on or before
the last date below.

Manufacturing Technology 2

P. J. Harris TEng(CEI), Associate Member IProdE, MIQA

Lecturer in the Department of Engineering, Woolwich College

The Butterworth Group

United Kingdom	**Butterworth & Co. (Publishers) Ltd** London: 88 Kingsway, WC2B 6AB
Australia	**Butterworths Pty Ltd** Sydney: 586 Pacific Highway, Chatswood, NSW 2067 Also at Melbourne, Brisbane, Adelaide and Perth
Canada	**Butterworth & Co. (Canada) Ltd** Toronto: 2265 Midland Avenue, Scarborough, Ontario M1P 4S1
New Zealand	**Butterworths of New Zealand Ltd** Wellington: T & W Young Building, 77–85 Customhouse Quay, 1, CPO Box 472
South Africa	**Butterworth & Co. (South Africa) (Pty) Ltd** Durban: 152–154 Gale Street
USA	**Butterworth (Publishers) Inc.** Boston: 10 Tower Office Park, Woburn, Mass. 01801

First published 1979

© Butterworth & Co. (Publishers) Ltd, 1979

British Library Cataloguing in Publication Data

Harris, P J
    Manufacturing technology 2.
    1. Manufacturing processes
    I. Title
    670   TS183   79-40369

    ISBN 0–408–00410–X

Typeset by Scribe Design, Medway, Kent
Printed by Page Bros Ltd, Norwich, Norfolk

Preface

As a result of the Haselgrave report and the recognition of the role of the technician in industry, the Technician Education Council (TEC) was formed. The courses offered by the TEC will now replace traditional courses such as ONC, HNC, and the City and Guilds mechanical engineering technicians. Since these courses have for a long time been regarded as technician qualifications, it seems logical that some rationalisation should take place to produce an overall technician qualification.

The style of the TEC courses is somewhat different in that conventional syllabuses are now replaced by subject units either of standard form as prepared by the TEC or college devised, designed around the needs of local industry. These units are based on a system of specified learning objectives, which indicate what the student has to know, rather than what the teacher has to teach. The Council have structured these units at various levels and the purpose of this book is to cover the work required for manufacturing technology at Level II of the A5 mechanical and production engineering programme. The unit for manufacturing technology, U76/056, has replaced an earlier unit, U75/035, as from January 1977. Since colleges with early submissions to the TEC will still be working to U75/035, it was felt necessary to include material to cover both units.

The technician working in a manufacturing organisation must have a good understanding of a wide range of manufacturing principles, sufficient for him to communicate with craftsmen, while at the same time being able to liaise and advise the technologist regarding the most economic methods of manufacture which will meet product design requirements.

While a subject of this nature must be considered from a practical point of view, some mathematical treatment must be considered, not just as an academic exercise, but to illustrate how scientific concepts may be applied to efficient and economic manufacture.

Throughout this book I have kept the topics in the same order as they appear in U76/056 and U75/035. The subject matter relating to the specific learning objectives have been dealt with in a more logical sequence compared with that indicated by these units. The Council do in fact say at the beginning of each of their standard units that these objectives are not in any particular teaching sequence.

At the end of each chapter I have included a number of short answer questions, which should assist the student in private study and also be of assistance to the teacher in the preparation of course work.

I would like to thank HMSO for allowing me to use some illustrations on presses from their 'Health and Safety at Work' publications. Thanks are also due to Mr V Betteley CEng MIMechE MIProdE who made helpful suggestions and comments during the preparation of this work. Also special thanks are due to Mrs Margaret Lunn who kindly gave up so much of her time to type the manuscript ready for publication.

Although this book is primarily directed to students studying manufacturing technology at level 2 of the TEC, it is hoped that others involved in engineering manufacture may gain something from the pages that follow.

P.J.H.

Contents

1 Heat treatment

INTRODUCTION

Steel in its various forms is perhaps the most widely used of all engineering materials, possessing physical properties mainly dependent upon chemical composition. Although these properties are superior to those of non-ferrous and plastic materials they may, however, be insufficient to meet the requirements of the component design specification. Alternatively these properties may have to be modified to ease manufacture of certain components. It is for these reasons that heat treatment is essential, and it is the only way that existing properties can be radically changed, e.g. to produce a steel of extreme hardness or softness.

Centuries ago heat treatment was regarded as being very much an art, practised mainly by blacksmiths and armourers, who didn't really understand why, for example, certain steels became hard when quenched. Nowadays, however, heat treatment is regarded as a precise science based upon established metallurgical principles, so that the engineer can predict and accurately control the change in physical properties for a given heat treatment process.

In this chapter, some of the important heat treatment processes relative to steel are described, together with the associated equipment required.

STRUCTURE OF STEEL

Steel is basically an alloy of iron and carbon, having a carbon content of up to about 1.5%.

If the structure of pure iron is examined, it will be found to consist of crystals of iron, chemically known as *ferrite*. When carbon is added to ferrite the structure of the iron changes considerably, producing a structure consisting of two constituents, one being ferrite and the other containing the carbon. This second constituent is known as *pearlite*; when viewed microscopically it appears in the form of dark portions with a background of ferrite. Pearlite is a soft, tough constituent. If this pearlite is viewed under higher magnification, it will appear as a laminated structure of ferrite and a chemical compound known as *cementite*. Cementite is extremely hard and is formed by the chemical combination of ferrite and carbon. Figure 1.1 illustrates these structures.

If the carbon content is further increased, then the formation of pearlite will also increase; if the steel contains 0.83% carbon then the structure will be entirely pearlite, since all the carbon has combined with all the ferrite. If the carbon content is increased above 0.83% an excess of cementite will be produced, which effectively means that there is more cementite in the steel than there is already contained in the pearlite (remember pearlite is a compound of ferrite and cementite). The resultant structure consists of pearlite surrounded by what is known as *free* cementite. This free cementite now replaces the ferrite in steels which have a carbon content less than 0.83%.

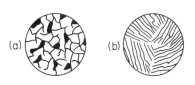

Figure 1.1 (a) 0.25% carbon steel, showing ferrite with small amounts of pearlite (dark areas) (×100 approx.), (b) pearlite under higher magnification showing alternate layers of ferrite and cementite (×1000 approx.), (c) structure of steel having carbon content in excess of 0.83%. Dark areas are pearlite surrounded by a network of cementite (×750 approx.)

IRON CARBON EQUILIBRIUM DIAGRAM

During heat treatment the structures so far described will undergo certain changes, and to understand what happens to these structures reference must be made to what is known as the *iron carbon equilibrium diagram*. This diagram is of considerable importance, since it relates the change in structure to the treatment temperature and the percentage carbon content. The complete diagram is somewhat complex and the one shown in Figure 1.2 represents a simplified version applicable to the heat treatment of plain carbon steels.

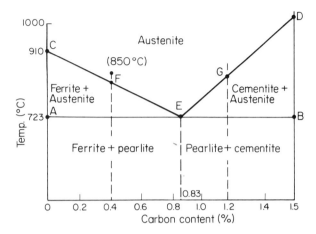

Figure 1.2 Iron carbon equilibrium diagram

There are two important temperature ranges associated with this diagram. The horizontal line AB (723 °C) is known as the lower critical temperature, while the sloping lines CE and ED are known as the upper critical temperature.

Consider now what happens in the case of a 0.4% carbon steel when it is heated from 0 °C to approximately 850 °C. Between 0 °C and 723 °C the structure will consist of ferrite and pearlite. As the temperature further increases a transformation takes place whereby the pearlite begins to change to form a new constituent known as *austenite*, which is the result of the carbon being dissolved into the ferrite. As the temperature increases beyond the point F (upper critical temperature for this steel) to 850 °C the structure is completely transformed to austenite, i.e. all the carbon has been dissolved by the ferrite. If the steel is now slowly cooled then the reverse will take place. As the temperature falls below F some austenite is changed back into ferrite and as the temperature drops further below 723 °C the remaining austenite will transform back into pearlite, resulting in a final structure of ferrite and pearlite.

Consider now the slow cooling of a 1.2% carbon steel which has been heated to a temperature above G (upper critical temperature) so that a completely austenite structure has been formed. As the temperature falls below G some austenite is transformed into cementite, until the lower critical temperature is reached, whereby the remaining austenite is changed to pearlite. The resultant structure will now consist of cementite and pearlite.

It will be noticed from the diagram that at 0.83% carbon the upper and lower critical temperatures coincide at the point E (723 °C) and as already stated this structure will be completely pearlite. A steel containing 0.83% carbon is known as a *eutectoid* steel. The word eutectoid is often used in metallurgy to denote a structure which is of laminated form. Remember pearlite consists of alternate layers of ferrite and cementite. A steel containing less than 0.83% carbon is referred to as a *hypo-eutectoid* steel, while one containing more than 0.83% carbon is known as a *hyper-eutectoid* steel.

So far we have only seen what happens to a specified plain carbon steel when its temperature changes. Let us now consider specific heat treatments and see how they are related to the iron carbon equilibrium diagram.

HEAT TREATMENT PROCESSES

In general there are four major heat treatments that may be applied to steels.

Annealing

Annealing is a general term used to describe a process whereby a steel is slowly cooled to change its properties for various reasons. There are three annealing processes commonly applied to steel.

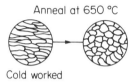

Anneal at 650 °C

Cold worked

Figure 1.3 Stress relief annealing

1. *Full annealing.* This is carried out to induce extreme softness and ductility, thus making the steel suitable for cold working. Cold working implies operations such as bending, rolling and drawing performed at room temperature. Steel castings are often fully annealed to improve their structure, since a cast structure is extremely coarse and irregular. Annealing will also relieve some of the stresses set up when the casting contracts due to solidification.

The process involves heating the steel to about 30 °C above its upper critical temperature for hypo-eutectoid steels, and by the same amount above the lower critical temperature for hyper-eutectoid steels. The steel is then maintained at this temperature to ensure that the internal changes take place (known as *soaking*) and then allowed to cool slowly within the furnace. This is best achieved by switching the furnace off, after soaking, so that the steel cools slowly as the furnace loses heat.

2. *Stress relief annealing.* When a steel is subjected to cold working it will work harden due to the severe deformation that takes place. To restore its former ductility so that further cold work can be performed, stress relief annealing is carried out. Stress relief annealing is sometimes known as *process annealing*. The process consists of heating the steel to a temperature of about 650 °C, followed by slow cooling.

Figure 1.3 shows the structure of steel after cold working and the structure that results after stress relief annealing.

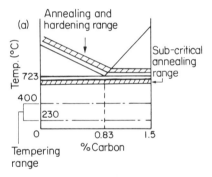

Figure 1.4 Heat treatment ranges related to the iron carbon equilibrium diagram

3. *Spheroidising.* This annealing process is used essentially to improve the machinability of high carbon steels, which can be difficult to machine due to the presence of large amounts of cementite. The process consists of heating the steel to a temperature between 650 and 700 °C for about 24 hours or more, and again followed by slow cooling. This prolonged heating causes the cementite layers within the pearlite to break up and reform as spherical globules. With the cementite now spheroidised machining will be found to be much easier.

Stress relief annealing and spheroidising are sometimes classified as *sub-critical* annealing processes since their treatment temperatures are below the lower critical temperature.

Figure 1.4(a) shows these annealing temperature ranges related to the iron carbon equilibrium diagram.

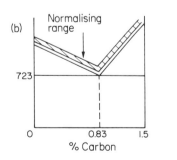

Normalising

The purpose of *normalising* is to produce a fine uniform grain structure, together with required mechanical properties.

The process consists of heating the steel to above its upper critical temperature, again by about 30 °C, followed by soaking. Cooling then takes place outside the furnace in still air, which produces a faster cooling rate compared with annealing, thus resulting in a finer grain size. As with annealing the soaking period will depend on the shape and dimensions of the component, but as a general rule it will be about 45 minutes for every 25 mm of thickness.

Compared with annealing, strength, hardness and toughness are increased but ductility is reduced. A normalised steel often gives a superior surface finish when machined.

Hardening

To *harden* a steel it should have a carbon content in excess of 0.4%, since less than this will not produce any worthwhile increase in hardness.

The process consists of heating the steel so that all or some austenite is formed, followed by rapid cooling, achieved by quenching. Quenching mediums used may be either brine, water or oil; these will be discussed later in this chapter. By cooling in this fashion pearlite is not allowed to form: instead a new structure is formed, known as *martensite*. When examined under the microscope, martensite has a fine needle-like structure, which is very hard and brittle. It will be noticed from Figure 1.4 that the annealing temperature range is also the same as the hardening range.

Tempering

A fully hardened steel consisting of martensite is extremely hard and brittle, making it susceptible to cracking and chipping when in service. For this reason it is usual to *temper* the steel, which will reduce its hardness slightly while at the same time its toughness is increased.

The process consists of reheating the steel below the lower critical temperature, between 230 and 400 °C, followed by quenching. This will cause some of the martensite to transform back into pearlite.

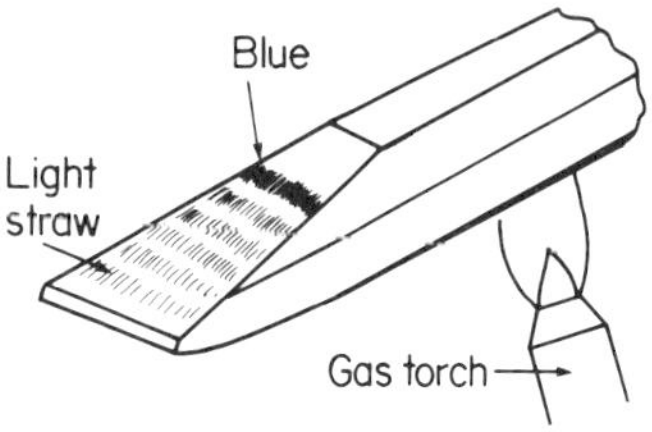

The exact temperature used will depend on the degree of temper required, but in general the higher the temperature the softer the steel becomes. This temperature may be indicated by the oxide colours formed (ranging from pale straw to dark blue) when a gas torch is played on to a polished part of the steel. Figure 1.5 illustrates this method for the tempering of a cold chisel.

The above method is regarded as a workshop technique, requiring a certain amount of skill, and has the disadvantage that colour perception will vary from one person to another. A better method is to use a furnace in conjunction with accurate pyrometric control. This will be dealt with more fully in the section on heat treatment equipment.

The table shows the tempering colours and temperatures together with typical applications.

Figure 1.5 Tempering a cold chisel

Temperature (°C)	Colour	Applications
230	Pale straw	Planing, slotting and turning tools, scrapers and hacksaws
240	Dark straw	Drills, reamers and milling cutters
250	Brown	Taps punches, dies and woodworking tools
260	Brownish purple	Plane blades and punches
270	Purple	Press tools, surgical instruments
280	Dark purple	Cold chisels, wood chisels
300–400	Blue	Wood saws, springs. For general toughening without exceptional hardness

QUENCHING

When steel is rapidly cooled by *quenching* sudden contraction takes place, and stresses are set up within the structure. If the correct technique is not adopted, distortion and quench cracks are likely to occur, which could mean the scrapping of what is otherwise good work. Long thin components are best quenched vertically end on, so that contraction takes place along the smallest dimension, thereby keeping the risk of cracking to a minimum. When the component has been immersed in the quenching medium it should be moved about, or agitated, so as to accelerate the cooling effect.

The severity of the quench will be dependent upon the medium used and will affect the properties required in the steel. The three most common quenching mediums used are:

(a) Brine   (b) Water   (c) Oil.

Brine will give the most rapid quench and, if the risk of cracking is to be avoided, should only be used on small components of uniform section requiring extreme hardness.

Water and oil are the most commonly used mediums. Water gives good hardness, but it can produce distortion if care is not taken. Oil is the least severe, best used on components where toughness is more important than extreme hardness. If water quenching is a problem due to distortion and cracking, then it is better to increase the carbon content of the steel and quench in oil. A high carbon steel quenched in oil can produce the same hardness as a steel having a lower carbon content quenched in water. The oil chosen for quenching should have a high flash-point, so that it does not ignite when it comes into contact with hot steel. In this respect oil baths should have a metal cover which can be used to extinguish any flare-up that might occur.

One final point is that the design of the component is extremely important if quenching is to be successful – a point that some designers seem to overlook. As a general rule all sharp corners and uneven sections should be avoided if possible. Figure 1.6 shows a simple component redesigned for quenching.

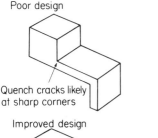

Poor design

Quench cracks likely at sharp corners

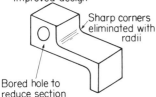

Improved design

Sharp corners eliminated with radii

Bored hole to reduce section

Figure 1.6 Design for quenching

EFFECT OF HEAT TREATMENT ON PHYSICAL PROPERTIES

The table shows a summary of the effect that the heat treatment processes so far discussed have on physical properties.

Heat treatment	Physical properties				
	Hardness	*Toughness*	*Ductility*	*Strength*	*Grain structure*
Hardening	Increased value depends on % carbon quenching medium	Decreased	Considerably decreased	Increased	Fine grain martensite
Tempering	Decreased slightly	Increased, depends on tempering temperature	No appreciable increase	Increased	Basically martensite with some re-formed pearlite
Annealing	Decreased	Increased	Considerably increased	Decreased	Ferrite plus pearlite of uniform size
Normalising	Slight increase	Increased	Reduced	Increased	Ferrite plus pearlite. More refined than annealing

HARDENABILITY

Hardenability may be simply defined as the depth to which a steel may be hardened. This is not to be confused with *hardness*, which is governed by the carbon content and to a certain extent the quenching medium.

If a steel is to be fully hardened it must be cooled rapidly from the austenite state to ensure a complete transformation to martensite. This, however, is not possible with plain carbon steels, since the cooling speed will decrease from the edge of the steel towards the centre. Effectively what has happened is that the mass of the steel has retarded cooling. If the structure is now examined it will be found to have an outer layer consisting of martensite and an inner section consisting of pearlite. When this occurs the steel is said to have poor hardenability. Figure 1.7 shows how the depth of hardness varies with the change in section.

The hardenability of a plain carbon steel may be improved by adding certain elements. A typical steel would have the following composition.

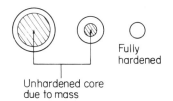

Fully hardened

Unhardened core due to mass

Figure 1.7 Effect of mass on hardenability

0.45% Carbon   1% Chromium
0.9% Manganese   0.2% Molybdenum

Such a steel is known as an *alloy* steel. Figure 1.8 shows the variation in hardenability for a plain carbon steel and an alloy steel as the result of taking hardness tests across the diameter of a test piece.

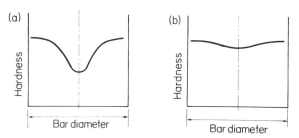

Figure 1.8 Hardenability curves, (a) plain carbon steel, (b) alloy steel

RULING SECTION

Although alloying elements improve hardenability it would be a mistake to assume that an alloy steel will harden right through irrespective of size of section. Because of this the steel manufacturer, in conjunction with the British Standards Institution (BSI), specifies the maximum diameter for which certain mechanical properties may be obtained, from a given heat treatment. This maximum diameter is known as the *ruling section*. The table shows values of tensile strength and hardness that can be expected from the ruling sections indicated, for a low alloy steel (BS970/608M38). It should be understood that if these sections are exceeded then the properties stated will not be obtained.

Ruling section (diameter (mm))	Tensile strength (N/mm^2)	Brinell hardness	Heat treatment
150	700	201	Oil harden
100	770	223	840/870 °C
63	850	248	and temper
30	930	269	550/680 °C

CASE HARDENING

It has been stated that a plain carbon steel having a carbon content of less than about 0.4% will show no appreciable increase in hardness when quenched. *Case hardening* is a method whereby such steels are given a carbon rich skin or case (0.8–0.9% carbon) which can be heat treated to produce a hard, wear resistant surface.

The process is divided into two parts:

1. Enriching the outer skin (case) with carbon, known as *carburising*.
2. Hardening the case and *refining* (toughening) the core.

The process of carburising may be carried out using either pack, liquids, or gaseous methods.

Pack carburising The components to be treated are packed into steel boxes, together with a suitable carburising material rich in carbon. The lids are then sealed with fire-clay, to exclude air and also to stop any gases from escaping. The boxes are then heated slowly to a temperature of 900–950 °C and held at this temperature for about 4–5 hours, depending on the depth of case required. It is during this time that the carbon from the packing material will penetrate into the outer layers of the components. Figure 1.9 shows the relationship between treatment time and depth of case for

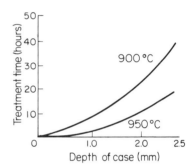

Figure 1.9 Relationship between depth of case and treatment time for a 0.15% plain carbon steel

temperatures of 900 and 950 °C. The carburising medium used consists of a mixture of charcoal, hoof clippings and bone dust – all substances rich in carbon.

After carburising the components are left to cool down within the boxes.

Liquid carburising Carburising by this method is particularly suitable for components that require a thin case of the order of 0.1–0.25 mm thick.

The process is carried out using baths of molten salt consisting of 20–50% sodium cyanide together with approximately 40% sodium carbonate with small additions of sodium chloride and barium chloride.

This mixture is heated in cast iron pots to a temperature between 850 and 950 °C, and using a suitable basket the work is immersed in the molten salt for a period of up to one hour.

Gas carburising The components to be carburised are placed in a gas-tight furnace and heated to a temperature of about 900 °C for 3 hours. The furnace atmosphere contains gases that will deposit carbon on to the surface of the work. These gases must be of a type known as *hydrocarbons*, which are gases consisting of hydrogen and carbon. Two such gases that are often used are methane and propane.

Gas carburising offers the following advantages:

1. The surface of the work is clean after treatment.

2. The carbon content of the surface layers on the component can be controlled more accurately, compared with pack or liquid carburising.
3. The technique is suitable for carburising large quantities.

Heat treatment after carburising

Whatever method of carburising is used, the prolonged heating at a high temperature will have made the grain structure at the core relatively coarse. This means that the core will have to be refined in addition to hardening the case. The core will still have a low carbon content (about 0.15%) which is refined by heating to just above its upper critical temperature (870 °C for a 0.15% carbon steel) followed by quenching in oil.

The case will have a carbon content between 0.8 and 0.9% carbon, which is hardened by reheating to just above the lower critical temperature (about 760 °C) followed by water quenching.

The component can now be tempered at about 200 °C to relieve any quenching stresses set up due to the hardening operation. The resultant structure will consist of a soft but tough core, with a hard, wear resistant case. Figure 1.10 shows these treatments in relation to the iron carbon equilibrium diagram.

Some components that are case hardened are required to have certain portions left soft, e.g. a screw thread. This may be achieved in one of three ways:

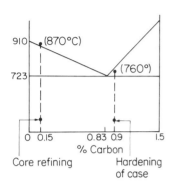

Figure 1.10 Case hardening related to the iron carbon equilibrium diagram

(a)   Leaving on excess metal, which after carburising may be machined off, so that the low carbon portion of the steel revealed will not be affected when the case is hardened.

(b)   Copper plating the areas not to be carburised, since carbon will not penetrate copper.

(c)   Coating the areas to be left soft with a paste of fire-clay.

HEAT TREATMENT EXAMPLES

The following examples will demonstrate the application of certain heat treatment processes.

Example 1

Figure 1.11 shows a lathe centre to be made from a 1% plain carbon steel. The main requirement of a centre is that it should be hard, to withstand the excessive friction imposed upon its tip.

After turning this centre should be hardened and tempered.

Figure 1.11 Lathe centre: 1% carbon steel

1. Since it is made from a 1% carbon steel (hyper-eutectoid), harden by heating to just above the lower critical temperature (about 750 °C) and water quench.

2. To remove stresses set up by hardening and to slightly induce a certain amount of toughness, temper at 230 °C.

After heat treatment the centre is then finished machined by grinding.

Example 2

Figure 1.12 shows a selector gear for use in a machine tool gearbox, which is to be capable of transmitting 1.9 kW at 2000 rev/min, continuously.

It should be apparent that this component will require to be wear resistant, and possess a high degree of toughness.

Since the component is supplied in the form of a forging the grain structure will be fairly coarse, and should be refined by normalising. This will also improve its strength and machining qualities. This involves heating to 900 °C (about 30 °C above the upper critical temperature)

Figure 1.12 Gear selector: 0.2% carbon steel forging

followed by a period of soaking. It is then removed from the furnace and allowed to cool in air.

The forging will now be machined, including the cutting of the teeth.

So that this component is to be tough and wear resistant, case hardening is recommended, and is carried out as follows.

(1) Assuming a case depth of 1.5 mm pack carburise at 950 °C for 5 hours.
(2) The core must now be refined and toughened, by heating to 860 °C followed by water quenching.
(3) The case is hardened by reheating to 760 °C followed by oil or water quenching.
(4) To relieve some of the quenching stresses induced in the hardening operation, the component may be tempered at about 200 °C.

BRITISH STANDARD SPECIFICATIONS

Before heat treatment is carried out reference should be made to the relevant British Standard specification, which lists the physical properties that can be expected from a given heat treatment. BS 970 : 1972 covers the specification and properties for plain carbon and alloy steel. This standard is also useful in evaluating any differences in properties that might occur during heat treatment, compared with that specified in the standard.

HEAT TREATMENT FURNACES

Furnaces are a very important piece of heat treatment equipment. Their function is to bring the component to the correct temperature, and remain at that temperature long enough to enable any internal changes in structure to take place. If a furnace is to be efficient, then this should be done economically with the minimum of heat loss.

The main types of furnaces used are gas fired, electric, salt bath and oil fired furnaces.

Gas fired furnaces

To produce sufficient heat, it is necessary to use a suitable burner, which operates on the well known Bunsen principle. Figure 1.13 illustrates this principle. If a combustible gas passes up a metal tube and is ignited at the top a relatively cool flame will be produced. The oxygen required for combustion is obtained from the surrounding air, known as secondary air. If, however, primary air is introduced at the bottom of the tube to produce a gas/air mixture then a hotter flame will result. In an industrial Bunsen burner this gas/air mixture is regulated by an adjustable shutter which enables the flame temperature to be controlled.

The simplest type of gas fired furnace consists of a box shaped enclosure lined with fire bricks, known as the *refractory lining*, and the outside is covered in a sheet steel casing.

The burners are located at the sides of the furnace, the flames being directed on to the arched roof providing reflected heat to the workpiece. A flue must be provided to exhaust the burnt gases.

Although this is a relatively simple furnace it does have the disadvantage that the products of combustion will affect the steel being treated. If excess air is supplied to the burners, then the products of combustion will be carbon dioxide, water vapour and oxygen, which will rapidly scale the steel and extract a certain amount of carbon. If excess gas is used then the products of combustion will be carbon monoxide, water vapour and some unburnt hydrocarbon (gases consisting of hydrogen and carbon). This time slight scaling occurs with some carbon deposited on to the surface of the steel. Clearly both these conditions are undesirable.

These problems may be overcome to a large extent by enclosing the

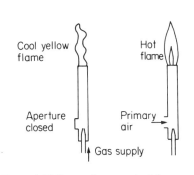

Figure 1.13 Bunsen burner principle

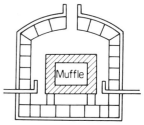

Figure 1.14 Gas fired furnace, (a) non-muffle furnace, (b) muffle furnace

work in a separate chamber known as a *muffle*, so that the hot gases do not come into contact with the steel. Figure 1.14 shows the construction of both muffle and non-muffle furnaces.

Although the muffle reduces scaling considerably, better protection can be obtained by introducing an inert gas into the muffle. Such a furnace is then known as a controlled atmosphere furnace. The atmospheres normally used consist of a mixture of hydrogen and nitrogen, which gives the steel an extremely clean surface, after treatment.

Electric furnaces   The basic construction of an electric furnace is similar to that of a gas fired furnace, with the exception that gas burners are replaced by electrical resistance heating elements, as shown in Figure 1.15. These are often in the form of a series of wire coils, very similar to that found in a domestic electric fire.

Electric furnaces offer the following advantages:

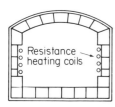

Figure 1.15 Electric furnace

(1)   No products of combustion present, hence no waste gases produced.
(2)   Clean in operation, and take up less space compared with gas fired furnaces of similar capacity (due to absence of pipework and ducting).
(3)   Lend themselves to automatic control of temperature.

The main disadvantage, however, is that at very high temperatures the heating elements tend to have a short life.

Salt bath furnace   This type of furnace basically consists of a cast iron pot containing molten salt, fitted into a cylindrical heating chamber, as in Figure 1.16. Heating may be by gas or electricity.

Because the work is completely immersed in the molten salt heating will be uniform, and since all air is excluded no scaling will be present.

Due to the dangerous nature of the salts used, certain safety precautions must be observed when using these furnaces:

(1)   Components should be pre-heated to ensure removal of oil and moisture, so that spitting is eliminated when work comes into contact with molten salt.
(2)   Food should not be consumed in the area where the salt bath is in use.
(3)   Because cyanide salts give off dangerous fumes, an efficient venting system is essential.
(4)   Heat treatment departments operating salt baths should carry an antidote for cyanide poisoning.

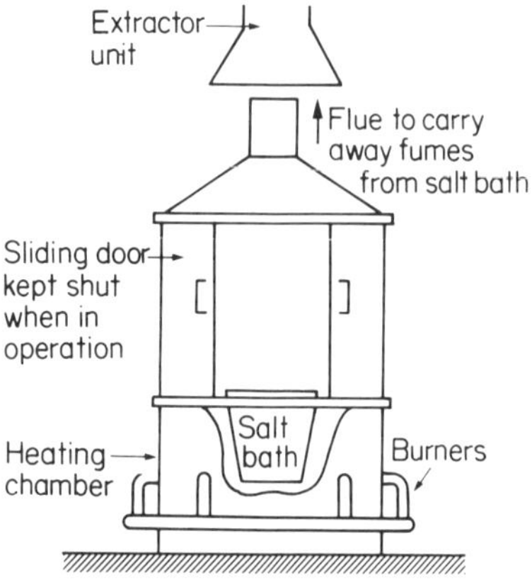

Figure 1.16 Salt bath furnace

Oil fired furnaces

Although gas and electricity provide the main source of heat for furnaces, fuel oil may also be used. Furnace construction is similar to that of the gas fired furnace, oil burners being used instead of Bunsen burners. Some furnaces are in fact designed so that they may operate on either fuels.

In general oil fired furnaces are not very common in heat treatment for the following reasons:

(1) Products of combustion present.
(2) Bulky oil storage equipment required.
(3) Oil has to be thinned down by pre-heating before entering burners.
(4) High cost of fuel oil.

TEMPERATURE CONTROL

It should now be apparent that temperature in heat treatment is extremely important, and should be controlled accurately if the required properties in the steel are to be obtained. It follows then, that some form of reliable temperature control must be available and ideally incorporated into the furnace construction. Devices used for this purpose are usually known as *pyrometers*. Although there are many, the two main types often used are the *mercury-in-steel thermometer*, and the *thermo-couple pyrometer*.

Mercury in steel thermometer

The familiar mercury-in-glass thermometer is not suitable for industrial heat treatment applications, since it is very fragile and may be difficult to read unless examined closely.

These problems are overcome by enclosing the mercury in a steel bulb which is connected to a Bourdon pressure gauge by means of a fine

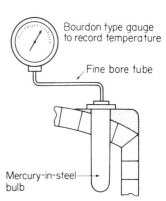

Bourdon type gauge to record temperature

Fine bore tube

Mercury-in-steel bulb

Figure 1.17 Mercury-in-steel thermometer

bore stainless steel tube, as shown in Figure 1.17. As the furnace temperature increases, the mercury in the bulb will expand and will be transmitted to the pressure gauge, calibrated to read in units of temperature.

The fine bore tube connecting the pressure gauge may be up to 30 m in length, which enables the gauge to be positioned for ease of reading. The maximum temperature that can be recorded with this system is about 600 °C.

Thermocouple pyrometer

The mercury-in-steel thermometer has the severe limitation that it can only measure temperatures up to 600 °C; most heat treatment temperatures are far in excess of this. Because of this thermocouple pyrometers are more frequently used, which can operate at temperatures as high as 1400 °C.

If two wires of dissimilar materials are joined together at their ends, and one end is at a higher temperature than the other, then a small

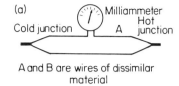

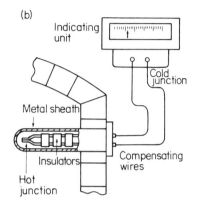

Figure 1.18 Thermocouple pyrometer

electric current will be produced in the circuit formed. The magnitude of this current will depend upon the difference in temperature between the hot and cold junctions. If a milliammeter is introduced into the circuit, then instead of recording units of current, it may be calibrated to read in units of temperature, since any increase in current will be the result of an increase in temperature of the hot junction. A device such as this is known as a *thermocouple.*.

In practice the indicating meter is made the cold junction, the connection to the furnace being made using compensating wires.

These wires are made from the same material as the thermocouple so as to compensate for any variation in temperature of the cold junction. Figure 1.18 illustrates the principle of a thermocouple together with a practical installation.

The temperature range that thermocouples may be subjected to will depend on the material from which the wires are made. The table shows some typical thermocouples.

Thermocouple		*Temperature range* (°C)
Copper	Constantan	−200 to 300
Iron	Constantan	−200 to 800
Chromel	Alumel	−200 to 1200
Platinum	Platinum/Rhodium	0 to 1400

Constantan has a composition	60% Copper	40% Nickel
Chromel has a composition	90% Nickel	10% Chromium
Alumel has a composition	95% Nickel	2% Aluminium
		3% Manganese
Platinum/Rhodium has a composition	90% Platinum	10% Rhodium

MECHANICAL TESTING

In carrying out heat treatment we are for various reasons changing the mechanical properties of the steel. So that these changes in properties meet the specification requirements of the manufactured component, some form of testing must be undertaken.

Strength

The *strength* of a material is its ability to resist either tensile or compressive forces without fracture.

Strength may be measured using a tensile testing machine in which a test piece is subjected to increasing loads until it finally fractures. Before the test piece is fitted to the machine two centre pop marks are made at a distance known as the gauge length. At these points an instrument known as an *extensemeter* is fitted which will record the extensions as the loads are applied. Figure 1.19 shows a typical load extension graph for a plain carbon steel in the normalised condition.

With reference to Figure 1.19 the applied load will be proportional to the extension between *AB* (Hooke's law), point *B* being known as the *elastic limit*. As the loads are further increased the steel will begin to yield at *C*, producing rapid extensions for relatively small increases in load. The point *D* on the curve represents the *maximum tensile load*. As the load increases further 'necking' or 'waisting' occurs, whereby the cross-sectional area is rapidly reduced, until the test piece finally fractures.

If the original cross-sectional area is known and the maximum load noted then

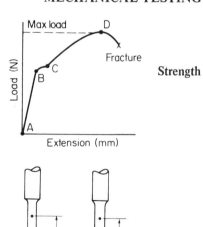

Figure 1.19 Tensile test

$$\text{Maximum tensile strength (N/mm}^2) = \frac{\text{Maximum load (N)}}{\text{Original cross-sectional area (mm}^2)}$$

The results obtained from a tensile test may also give an indication of the steel's ductility, by considering the amount that the gauge length extends. This extension is usually referred to as the *percentage elongation*. The higher this figure the greater the ductility.

$$\text{Percentage elongation} = \frac{\text{Extension of gauge length}}{\text{Original gauge length}} \times 100$$

Hardness   This is the property of a material to withstand indentation, scratching, and wear by other materials.

There are a number of methods by which hardness can be measured; four common tests are the Shore scleroscope, Brinell, Vickers, and Rockwell.

SHORE SCLEROSCOPE
This is a small portable instrument, ideally suited for the testing of heavy components which are not easily transported, e.g. large forgings and machine beds. The instrument is shown in Figure 1.20.

The test consists of dropping a diamond tipped hammer of mass 2.6 g from a height of 250 mm within a glass tube graduated into 140 equal divisions. The height of rebound is a measure of hardness: the harder the material the higher the rebound.

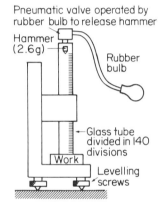

Figure 1.20 Shore scleroscope

BRINELL TEST
This test, depicted in Figure 1.21, uses a hardened steel ball indenter which is forced into the surface being tested, using a steady load. The diameter of impression made is then measured using a low power microscope. The diameter of ball usually used is 10 mm and the load used for steel is 3000 kg. The *Brinell hardness number (B.H.)* is given by

Figure 1.21 Brinell hardness test

$$B.H. = \frac{\text{Load } (F)}{\text{Surface area of impression}} = \frac{F}{\pi D/2[D - \sqrt{(D^2 - d^2)}]}$$

where D = diameter of ball used and d = diameter of impression.

To overcome tedious calculations, *B.H.* may be obtained directly from specially prepared tables that relate measured impression to the load and ball diameter used.

This test has the disadvantage that the ball indenter will deform under load when testing very hard steels.

VICKERS PYRAMID TEST
This test, shown in Figure 1.22, is similar to the Brinell test except that a diamond indenter is used, in the form of a square based pyramid having an apex angle of 136°. After the indenter is removed, the impression is measured across the diagonals using a low power microscope. The reading obtained is then referred to tables, relating size of impression and the load used, to find the *Vickers pyramid number*.

The advantage of this test compared with the Brinell test is that it is suited to much harder materials. Both tests can often be carried out on the same machine.

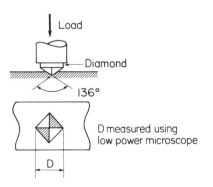

Figure 1.22 Vickers pyramid test

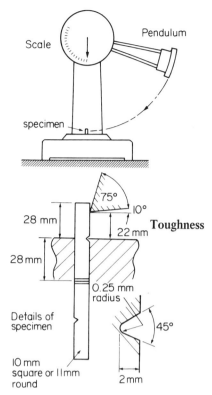

Figure 1.23 Izod impact test

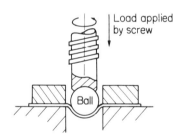

Figure 1.24 Erichsen ductility test

ROCKWELL TEST

Unlike the Brinell and Vickers test, the Rockwell method measures hardness using the *depth of penetration* of an indenter. This indenter is in the form of a hardened steel ball (1.5 mm diameter) or a conical shaped diamond.

A minor load of 10 kg is first applied to give an initial penetration and also to take out any slackness in the machine. A major load is then applied 150 kg for the diamond and 100 kg for the steel ball, and the depth of penetration recorded on a dial to give the Rockwell hardness number. The dial has two scales: scale *B* is used for the steel ball and scale *C* for the diamond.

This method is found to be suitable for rapid routine hardness testing on a production basis.

Toughness   *Toughness* is the ability of a material to withstand impact and shock loads. A tough material will absorb a considerable amount of energy before it fractures.

A common test to measure toughness is the *Izod impact test*, shown in Figure 1.23. In this test a notched specimen is held in a vice and a heavy pendulum allowed to strike the specimen from a fixed height. The energy at impact is 163 J and the amount of energy required to fracture the specimen will be an assessment of its toughness. This value is recorded by a pointer which comes to rest when the pendulum reaches the highest point of its swing.

It is usual to repeat this test three times on the same specimen and take an average value.

Ductility   A *ductile* material is one which can be drawn or rolled out into long lengths without fracture.

Although the percentage elongation obtained from the tensile test gives a measure of ductility, the *Erichsen cupping test* (Figure 1.24) is another method designed to assess ductility of steel in sheet form. A specimen is clamped between a die and blank holder and a 20 mm diameter hardened steel ball forced into the specimen, until it just begins to split. The depth of cup formed represents the *Erichsen ductility value*.

One additional feature that this test shows is the grain structure of the specimen. If the dome is rough and crinkled, then in steel this indicates a coarse structure, possibly due to insufficient annealing after cold work. If the steel has been normalised, then the dome should be smooth.

SUMMARY   Heat treatment in steel is carried out to modify its internal structure, so that its mechanical properties are improved to meet certain specified requirements. The iron carbon equilibrium diagram is an important aid in predicting what structural changes take place in a plain carbon steel, and also determines the treatment temperature given the percentage carbon content.

Annealing and normalising involve heating, followed by slow cooling. Annealing renders the steel soft and ductile, while normalising refines the grain structure and improves machinability. Hardening is carried out by rapid cooling from an elevated temperature, followed by tempering to relieve quenching stresses and to toughen steel. Quenching mediums frequently used are brine, water and oil. The depth to which a steel may be hardened is known as its hardenability, and the maximum properties that can be expected from a hardening and tempering operation are controlled by its ruling section.

If a low carbon steel (0.4% C or less) is to be hardened, then it can only be case hardened, which will produce a hard, wear resistant surface together with a soft but tough core.

Furnaces are used to raise the temperature of the steel, which may use gas, electricity or oil as a source of heat. Care must be taken when using salt bath furnaces, due to the dangerous nature of the salts used. Muffles and controlled atmospheres are used in gas fired furnaces to protect the steel from the products of combustion. Pyrometers are used to control furnace temperatures, the type used being dependent upon the magnitude of the temperature to be measured.

Mechanical testing is an important aspect of heat treatment. Tests are commonly carried out to measure hardness, strength, toughness and ductility.

QUESTIONS

(1) Construct a simplified iron carbon equilibrium diagram between 0–1000°C and 0–1.5% carbon.

(2) Explain the meaning of the terms:
(a) pearlite
(b) cementite.

(3) With the aid of the iron carbon equilibrium diagram show the heat treatment ranges for the following:
(a) Annealing
(b) Normalising
(c) Hardening.

(4) State two reasons for normalising a plain carbon steel.

(5) Name two factors that determine the hardness of a steel during heat treatment.

(6) Give two reasons for tempering. What is the approximate temperature range for tempering a plain carbon steel.

(7) What is meant by the term 'carburising'?

(8) Explain why it is necessary to refine the core during a case hardening operation.

(9) Explain what is meant by the term 'hardenability'.

(10) Explain what is meant by the term 'ruling section'.

(11) Show by means of a graph how the hardenability of a plain carbon steel compares with that of an alloy steel.

(12) State three quenching mediums, in order of hardness produced on a plain carbon steel.

(13) Give two possible reasons for cracking in the hardening of a component.

(14) What is the purpose of a muffle in a gas fired furnace?

(15) Explain what is meant by a controlled atmosphere furnace.

(16) State two advantages for using electricity as a source of heat for a heat treatment furnace.

(17) Explain the principle of the thermocouple pyrometer.

(18)   State two advantages of using a thermocouple pyrometer compared with a mercury-in-steel thermometer.

(19)   State two safety precautions that should be observed when using a salt bath furnace.

(20)   Describe a common test that may be used to measure two of the following properties:

   (a) Strength   (b) Ductility
   (c) Hardness   (d) Toughness.

2 Plastics

INTRODUCTION   The term *plastic* is a general term, used to describe a group of non-metallic materials that at some stage in their life were capable of plastic flow, before being shaped to form the final component.

The application of plastic materials has rapidly increased over the last few decades, replacing components that were traditionally made from metal. Despite this increased use of plastics, it is doubtful at present that they will completely replace metals, since they are not suitable for parts that are to be subjected to high stresses, wear and elevated temperatures. Yet it is interesting to note that the Royal Navy have now in commission a mine-sweeper whose hull, deck and superstructure are constructed entirely from glass reinforced plastic.

It is unfortunate that cheapness is sometimes equated with plastic products. This is often an unfair statement since the product in question probably should not have been made from plastic in the first place! By careful design and material selection, plastics can offer properties which are compatible if not superior to those of some metals, e.g. they are extremely light and offer good protection to atmospheric and chemical attack.

The chemistry of plastics is extremely complex, being a technology in its own right, and one which is constantly developing. However, it is not the intention of this chapter to deal with the chemistry of plastics, but to consider the various moulding processes commonly found in the manufacture of plastic products.

PLASTIC MATERIALS   Plastics have been grouped into two main types depending on how they react to heat: *thermoplastics* and *thermosetting plastics*.

Thermoplastics   Plastics in this group may be softened by the application of heat and allowed to harden by cooling. This may be repeated indefinitely providing the heat does not cause damage.

Examples of thermoplastics are:

Nylon
Polypropylene
Polystyrene
Polyvinyl chloride (PVC).

Thermosetting plastics   Plastics in this group will undergo an irreversible chemical change during hardening. This means that they cannot be softened by heat for further application.

Examples of thermosetting plastics are:

Epoxy resin
Phenol formaldehyde (bakelite)
Polyester
Silicone.

RAW MATERIAL FOR MOULDING

The plastics stated previously are seldom used on their own, but are often blended with other constituents for specific reasons. These constituents are:

(1) *Filler*, used mainly to increase strength, but may be added for other reasons, e.g. to improve electrical insulation properties. Typical fillers used include wood dust, cotton flock, metallic oxides, asbestos and glass fibre. The filler can represent up to as much as 50% of the finished product.

(2) *Plasticiser*, used to soften the plastic, to put it in a better condition for moulding.

(3) A *hardening additive*.

(4) An *accelerator*, used to speed up the effect of the hardening additive

(5) *Colouring additive*.

The basic plastic material is known as the *binder*.

All the above constituents are added by the plastic manufacturer, who would control the composition, depending on the specification of the plastic.

The form in which the raw material is supplied before moulding is either in the form of pellets (about 3 mm × 3 mm), granules or powder.

MOULDING METHODS

Moulding represents by far the most common method of shaping plastic, since it is possible to shape the component directly from the raw material. In many cases the finished moulding will not require any further work such as machining.

In general there are four main types of moulding methods: injection, compression, transfer and blow moulding.

Injection moulding

Injection moulding is mainly confined to thermoplastics, and is considered to be the most important moulding technique, due to its suitability for volume production.

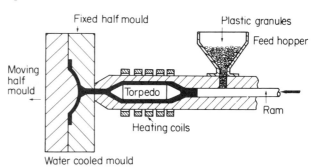

Figure 2.1 Ram injection moulding machine

Figure 2.1 shows the principle of the *ram injection* moulding machine. Plastic granules are metered into the injection cylinder from a hopper, and then compressed into the heating chamber by the ram. As the ram moves forward, the molten plastic from the previous charge is forced into the mould cavity. The purpose of the *torpedo* is to reduce the thickness of plastic adjacent to the heating chamber walls, since plastics are poor conductors of heat. After injection, the moulds are split and the component ejected, often assisted by ejector pins. The main disadvantage of the ram machine is that heating tends to be uneven.

Due to the development of plastics and moulding techniques, the ram

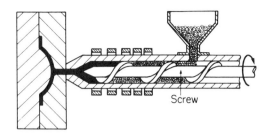

Figure 2.2 Screw plasticiser injection moulding machine

machine has largely given way to the *screw plasticiser* type as shown in Figure 2.2.

This machine uses a rotating screw which pushes the plastic forward into the heating chamber, and also feeds the molten plastic into the mould. Although electrical heating coils are used, the mechanical energy generated by the screw also contributes to the total heating effect.

The temperatures and pressures required for injection moulding range from 150 to 300 °C and 100 to 150 N/mm² respectively.

The process offers the following advantages:

1. Machines may be automatic or semi-automatic using unskilled labour.
2. Fast moulding times: for a medium sized component, 40 seconds is a typical figure, of which 30 seconds would be allowed for cooling, before the mould is split.
3. A wide range of plastics may be moulded.

Compression moulding This method is mainly confined to the moulding of thermosetting plastics, and to a limited extent thermoplastics that are too large to be injection moulded.

In this process two heated split moulds are mounted in a mechanical or hydraulic press. A measured amount of plastic, usually in the form of pellets, is placed into the lower mould. The press is then closed, thus bringing the two moulds together and, in so doing, the plastic softens due to the heat, and is formed to shape by the pressure applied. The component is then left in the mould to cure (harden), which may be of the order of a minute or more, thus making compression moulding a slow process compared with injection moulding. Its main advantage, however, is that comparatively cheap tooling is used, and that large mouldings of simple design may be manufactured by this method.

The temperatures and pressures used are within the range 120 to 190 °C and 20 to 70 N/mm² respectively.

Figure 2.3 shows the design of a typical press suitable for compression moulding.

The moulds used for compression moulding are basically of two types: *open flash* mould, and *closed* mould.

In the open flash mould, the lower mould is charged with a measured amount of plastic material, and the two moulds are brought together making contact at A, as shown in Figure 2.4(a). Excess plastic will then flow into the gutter provided, thus forming a flash (or fin) on the moulding, which is later removed.

It is important with this type of mould that guide pins are provided so as to maintain the positional relationship between the upper and lower moulds.

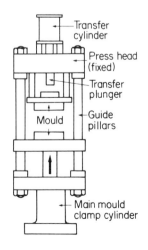

Figure 2.3 Compression and transfer moulding press

In the closed mould technique no flash is produced. This means that the plastic charge must be carefully controlled, since insufficient or excess plastic will make the moulding under- or over-size. The other advantage, apart from the absence of any flash, is that the pressure

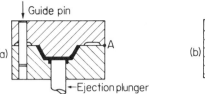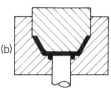

Figure 2.4 Compression moulds, (a) open flash mould, (b) closed mould

available from the press is exerted on to the moulded component; conversely, with the open flash mould some pressure is distributed on to the flash itself. Figure 2.4(b) shows the design for a closed mould.

Transfer moulding

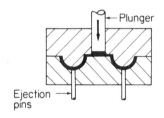

Figure 2.5 Transfer mould

In *transfer moulding*, the plastic material in the form of pellets is placed into a separate chamber and then transferred by a plunger into a heated closed mould. The main attraction of transfer moulding is that there is little pressure inside the mould cavity until it is completely filled. This offers the following advantages:

1. Delicate and intricate sections may be moulded without risk of distortion.
2. Inserts are unlikely to be misplaced due to inrush of molten plastic. (The use of inserts will be discussed later.)

The process also lends itself to the production of several mouldings simultaneously, positioned around a central plunger.

Figure 2.5 shows a typical transfer mould.

Blow moulding

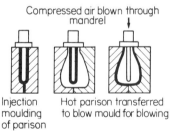

Figure 2.6 Blow moulding

This process is used for producing thermoplastic products in the form of bottles and other hollow components, which may not be suitable for other moulding methods. The process involves injection moulding a plastic blank, known as a *parison*, around a *mandrel*. While it is still hot, the parison is transferred to a *blow mould*, where compressed air is passed through the mandrel, resulting in the plastic taking up the shape of the mould cavity.

The threaded neck of bottles may be moulded during the injection moulding of the parison.

Figure 2.6 shows the stages in moulding a plastic bottle.

MOULDED INSERTS

It is common for some mouldings to have features such as tapped holes, keyways or splines incorporated into their design. Although these features can be machined into a moulding it is often more practicable and economic to produce them in the form of a metal insert. These inserts are secured in the mould in such a way that their positions are not altered due to the inrush of molten plastic. Provision must be made whereby the insert is firmly attached to the moulded plastic. Figure 2.7 shows various ways in which this may be achieved. It is desirable that inserts have coefficients of expansion similar to the plastic. If the contraction of the

Figure 2.7 Moulded inserts

plastic is greater than that of the insert, then internal stresses will be set up in the plastic when it hardens, which could lead to cracking during the moulding's life. A further consideration is that the metal chosen for the insert must not react chemically with the plastic. A brass insert, for example, will react chemically when moulded into polyvinyl chloride (PVC).

MOULD DESIGN

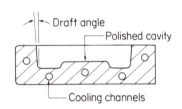

Figure 2.8 Mould design

In general the design and manufacture of moulds represent a very high standard of precision engineering, the cost of which are dependent upon the quality and complexity of the component to be moulded. Whatever the cost the mould will represent a considerable investment in the manufacturer's moulded product, since moulds are very expensive to design and manufacture. Moulds are made from hardened steel (sometimes case hardened), although mild steel or cast iron may be used if small quantities are to be moulded and wear is not an important consideration. The internal cavity is machined to a high standard of accuracy, and is finished by polishing, since any minute defect or scratch would be reproduced in the finished moulding. To facilitate removal of the moulding draught angles must be provided as shown in Figure 2.8. Further assistance may be obtained using ejector pins, which must be carefully positioned so that the moulding is not distorted when it is ejected. Moulds used for injection moulding of thermoplastics are water cooled, whereby water is permitted to flow through a system of holes within the mould body.

CHOICE OF MOULDING PROCESS

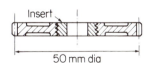

Figure 2.9 Nylon spur gear, (a) cross-section, (b) gears moulded on to a tree

Before choosing a moulding process the following factors must be taken into account:

1. Type of plastic to be moulded
2. Size of component
3. Quantity to be manufactured
4. Cost of tooling required.

The following example will demonstrate how a moulding process may be justified.

Figure 2.9 shows a spur gear to be moulded in nylon at a rate of 240 per hour. Since nylon is a thermoplastic and a large quantity is to be produced, then the most suitable process would be injection moulding. We must now consider if this process can meet the target figure of 240 per hour. The total moulding time for this component would be about 30 seconds, which represents 120 components per hour. To achieve our figure of 240 we can simply have two identical cavities in the same mould so that the gears will be ejected on a *tree*. Although producing two components from the same mould increases the mould cost, it is justified since the moulding cost attributed to the component is effectively halved.

SAFETY   Because the moulding of plastics requires the use of elevated temperature and pressures, personnel operating moulding machines must wear protective clothing. Hot molten plastic can produce very unpleasant burns. Caution should be exercised when handling raw plastic material, since some personnel may be susceptible to skin irritation (remember that plastic is a complex chemical compound). Obviously the wearing of protective gloves is the solution. All moulding machines must be properly guarded. The guard system should be so designed that the moulds will not close until the guard is in position, thus eliminating the possibility of fingers, arms or clothing becoming trapped. This is particularly important on injection moulding machines where the opening and closing of moulds may occur many times per minute.

SUMMARY   Thermoplastics may be softened by the application of heat, while thermosetting plastics cannot since they harden by chemical action.

The raw material used for moulding plastics consists of various constituents to improve physical properties and moulding characteristics, the basic plastic material being known as the binder.

Injection moulding is used mainly for thermoplastics and is capable of high outputs. Compression moulding is a slower process confined mainly to thermosetting plastics. Transfer moulding is used to mould components of intricate shape in thermosetting plastic, or where inserts are lightly secured, since moulding pressures are low.

Tapped holes, bushes, splines etc. may be produced in a moulding in the form of metal inserts. These are positioned in the mould prior to moulding.

Moulding machines must be properly guarded, and operating personnel should wear protective clothing.

QUESTIONS

(1)   Explain the difference between:
(a)   A thermoplastic
(b)   A thermosetting plastic.

(2)   State three forms in which raw plastic material can be used for moulding. Name two constituents found in the raw material and explain their function.

(3)   Describe, with the aid of sketches, the principle of:
(a)   Injection moulding
(b)   Compression moulding.

Compare the main advantages and limitations of these two processes.

(4)   Describe, with the aid of a sketch, the principle of transfer moulding, stating its main area of application.

(5)   For two components of your choice show how moulded inserts may be used to advantage. In each case indicate clearly how the insert is secured to the plastic.

(6)   Name three design features that should be incorporated into the design of a mould suitable for plastic moulding.

(7)   Compare the use of the open flash and closed moulds for compression moulding.

(8)   Describe what factors should be considered before a moulding process can be justified.

(9)   Describe, with the aid of a sketch, the process of blow moulding, and state what type of product this process is suitable for.

(10)   Name two sources of danger associated with plastic moulding machines, and state the relevant precautions.

3 Presswork

INTRODUCTION

The term *presswork* is here used to describe a range of processes which are performed on thin metal strip or sheet, mainly of low carbon steel and non-ferrous metals such as brass, copper and aluminium alloys. The thickness of material used seldom exceeds 3 mm.

Components can be produced rapidly using press tools and, unlike machining operations, very little waste is produced. Press tools, except for very simple tools, are expensive to design and manufacture, which means that presswork is mainly confined to large scale production. For example, a press tool used to form a car body panel will cost many thousands of pounds, but this cost is easily recovered since these panels are produced in large quantities.

In general, presswork makes a significant contribution to the total output of manufactured products, and in this chapter some of the more common presswork operations are described, with special emphasis made on the safe working of both hand and power presses.

PRESSWORK OPERATIONS

There are many presswork operations and these may be classified as being either shearing or forming processes. The table indicates how these processes relate to specific operations.

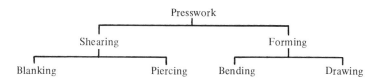

BLANKING AND PIERCING

A *blanking* operation refers to the shearing, by punching out, from flat material. The part removed, or *blank*, is the part that is required. This is often a first operation, whereby the blank may require further working, e.g. bending.

A *piercing* operation refers to the punching, i.e. shearing of holes of any shape in strip material, or in previously sheared blanks.

Figure 3.1 shows the essential elements of a press tool designed for blanking and piercing.

In blanking the die is made to size and the clearance subtracted from the punch, and in piercing the punch is made to size and clearance added to the die. The exact value of the clearance will depend on the nature of the material, but in general will be between 5 and 10% of the material thickness.

Punches and dies

Punches and *dies* are made from an alloy steel having a high carbon and high chromium content, hardened and tempered. The advantage of using an alloy steel is that distortion is minimised during heat treatment. It will be noticed (Figure 3.1) that the die is backed off by about $3°$; this

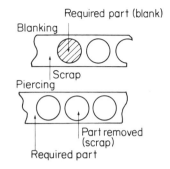

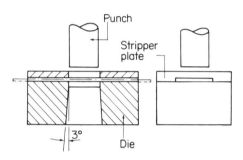

Figure 3.1 Basic elements of a press tool designed for blanking and piercing

is to ensure that the sheared blanks do not jam, but fall easily into the collecting basket.

A device known as a *stripper plate* is provided so that when the tool is completing its stroke it does not cling to the strip material.

Force required for blanking and piercing

To ensure that the press can provide a sufficient shearing force, it is often necessary to calculate the force required for a specified operation. This force will depend on the maximum shear strength of the material and the area to be sheared.

In general

$$\text{Force} = \text{Stress} \times \text{Area}$$

Therefore

$$\text{Punch force} = \text{Maximum shear strength} \times \text{Area to be sheared}$$

Note that the shear area will be the material thickness multiplied by the perimeter to be sheared.

Dimensions in mm

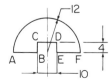

Figure 3.2

Example. The component shown in Figure 3.2 is to be blanked from steel strip 1.5 mm thick. Determine the punch force required, if the maximum shear strength of the material is 430 N/mm^2.

$$AB = EF \quad \text{and} \quad CB = DE$$

Therefore

$$\text{Perimeter to be sheared} = \pi r + 2(AB + CB) + CD$$
$$= (\pi \times 12) + 2(7 + 4) + 10$$
$$= 37.7 + 22 + 10$$
$$= 69.7 \text{ mm}$$

Therefore

$$\text{Punch force} = \text{Maximum shear strength} \times \text{area to be sheared}$$
$$= 430 \times 69.7 \times 1.5$$
$$= 44956.5 \text{ N}$$

say

$$\underline{45 \text{ kN}}$$

Shear on tools　In order to reduce the force required for blanking and piercing, especially on components of large thickness and/or complex profiles, shear may be applied to either the punch or die. For blanking, shear is put on the die since the flat face of the punch will produce a flat blank. For piercing, shear is put on the punch so that the waste material is

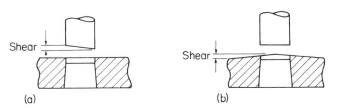

Figure 3.3 Shear applied to press tools, (a) single shear on punch, (b) double shear on die

deformed, thus ensuring that the strip is held flat against the die face. Shear may be single or double; the latter is to be preferred, since it tends to neutralise side thrust on punch or die. Figure 3.3 shows examples of shear.

Blanking layouts　Before a press tool for blanking is designed, a blanking layout (sometime known as a strip progression) must be produced, so that the maximum utilisation of material is obtained. It must be remembered that presswork is used for high volume production, so that a small saving on one component will represent a considerable saving on large quantities.

As a general rule the distance between the nearest point on the blanks and the distance between blank and edge of strip should not be less than the strip thickness. Some blanks may require bending at a later stage. This must be borne in mind when designing blank layouts, since the strip grain should be at right angles to the proposed bend. Figure 3.4 shows various ways in which a blank may be positioned on a strip.

Figure 3.4(b) represents the most economical use of material, whereby two punches may be used, or one punch used and strip fed through twice Although a press tool using two punches is more expensive to produce,

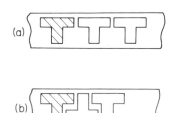

Figure 3.4 Blanking layouts

this increased cost is easily justified because of the increased output from the tool.

STANDARD DIE SETS   In any presswork operation it is important that the punch and die are accurately aligned. This is achieved using a *standard die set* consisting of a top and bottom bolster, which move relative to each other on guide pillars. During the manufacture of a press tool, the punch is fitted to the

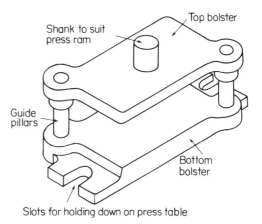

Figure 3.5 Standard die set

top bolster and the die to the bottom bolster. This enables the whole assembly to be fitted to the press, the alignment having been made during the manufacture of the tool. Figure 3.5 shows a standard die set.

PROGRESSION OR   Many components produced by press tools often require both blanking
FOLLOW ON TOOL   and piercing operations to be performed on them. This may be achieved by using a *progression tool*, where the strip is fed through and progressively worked on. Figure 3.6 shows the manufacture of a washer using

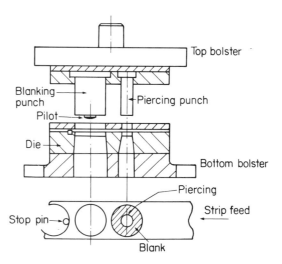

Figure 3.6 Follow on pierce and blank tool

such a tool. The hole is first pierced, then the strip is moved forward to be blanked, while at the same time the next washer for blanking is being pierced. It will be noticed that a pilot is fitted to the blanking punch. This will locate in the previously pierced hole, so that concentricity between hole and blank is maintained.

BENDING   Material used for bending should be ductile and, as already stated, the grain should be at right angles to the direction of bending. If bending takes place along the grain then cracking is likely to occur. The term *grain* is used in much the same way as when applied to timber. If the surface of sheet metal is examined closely, a series of parallel lines will be observed, as evidence of the direction of working by cold rolling, thus giving the material this so called grain characteristic.

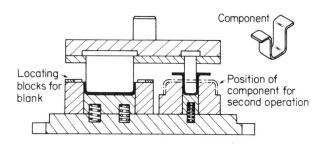

Figure 3.7 Two stage bending tool

Figure 3.7 shows a bending tool for producing the component shown. This component is being formed in two stages.

The blank for bending is located on the die. As the punch descends the blank is held against the punch by means of a pressure pad. When bending is complete, this pad will also eject the component from the die.

When a vee bend is to be produced on a press tool, as shown in Figure 3.8, the punch and die must be made slightly less than the included angle of the vee. This is to allow for the natural elasticity of the material to open out the vee, when the component is ejected. This is often known as *springback*. For a 90° bend, the punch and die would be made to an angle of say 88°. The exact figure will be dependent upon the elasticity and thickness of material being formed.

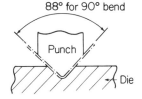

Figure 3.8 Springback

Allowance for bending   When a material is subjected to bending the outer layers of the bend will be in tension, causing the material to increase in length. The inner layers will thus be in compression and will cause the material to shorten. This effect produces a characteristic deformation on the edge of the material, which the reader will no doubt be familiar with when bending a metal strip in a vice. There will be somewhere between these inner and outer layers a condition where the material is neither in tension or compression. This will occur along an imaginary line known as the *neutral axis*. Although the neutral axis of the undeformed material is at half the material thickness it will, however, shift towards the compressed side during bending. So that the correct blank length is obtained prior to bending, the length of the curved portion *AB* in Figure 3.9(b) along the neutral axis must be determined. This is known as the *bend allowance* and is added to the straight portions of the component, enabling the final blank length to be established.

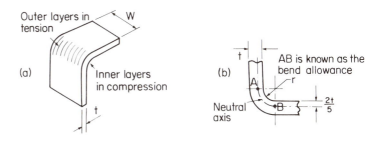

Figure 3.9 Allowance for bending

The bend allowance may be obtained from the following expression:

$$\text{Bend allowance} = 2\pi(r + \tfrac{2}{5}t) \times \frac{\theta}{360}$$

where r = radius of inside bend
t = material thickness
θ = angle of bend

Force required for bending   The punch force required for bending is governed by the bending area and the bending strength of the material. In the majority of cases the bending strength is assumed to be half the shear strength.
Therefore

$$\text{Punch force} = \text{Bending area} \times \frac{\text{Shear strength}}{2}$$

$$= W \times t \times \frac{\text{Shear strength}}{2}$$

where W = width of material
t = material thickness.

Note that if more than one bend is to be formed simultaneously, then for the purpose of the calculation W will equal the width of material multiplied by the number of bends.

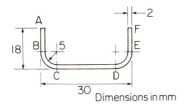

Dimensions in mm

Figure 3.10

Example. The component shown in Figure 3.10 is to be formed from a blank 25 mm wide. If the shear strength of the material is 430 N/mm² determine:
(a)   The developed blank length
(b)   The punch force required.
(a) $AB = EF$

Total length of straight portions $= 2AB + CD$

$$= 2(18 - 7) + (30 - 14)$$

$$= 22 + 16$$

$$= 38 \text{ mm}$$

Length along single bend $= 2\pi(r + \tfrac{2}{5}t) \times \dfrac{\theta}{360}$

$$= 2\pi(5 + \tfrac{2}{5} \times 2) \times \frac{90}{360}$$

$$= 2\pi \times 5.8 \times \tfrac{1}{4}$$

$$= 9.11 \text{ mm}$$

Therefore,

Total developed length $= 38 + (2 \times 9.11)$

$\qquad\qquad\qquad\qquad = 38 + 18.22$

$\qquad\qquad\qquad\qquad = \underline{56.22 \text{ mm}}$

(b)   Punch force $= W \times t \times \dfrac{\text{Shear Strength}}{2}$

Since there are two bends, $W = 2 \times 25 \text{ mm} = 50 \text{ mm}$

Therefore,

$$\text{Punch force} = 50 \times 2 \times \frac{430}{2}$$

$$= 100 \times 215$$

$$= 21500 \text{ N}$$

$$= \underline{21.5 \text{ kN}}$$

DRAWING

The function of *drawing* is to produce components of dish or cup shape from a flat sheet metal blank. Materials used for drawing should be ductile since during the process the material will undergo severe plastic flow. Some materials may have to be annealed to put them into the best condition prior to drawing.

Figure 3.11 shows a typical drawing tool for producing a cylindrical container. The blank is gripped by a punch and plunger, and as the punch descends the blank is drawn over the die and around the punch, thus forming a cup. Due to the severe compressive stresses induced in the material as it is drawn over the die, wrinkling is likely to occur. This is prevented by the pressure pad which keeps the blank rim flat as the draw is completed. It will be apparent from Figure 3.11 that as the draw gets deeper then the pressure on the blank rim increases, since the compressive stresses will also increase. Care must be exercised when regulating the pad pressure, since if it is too great the material may tear.

A rubber block applies pressure to the plunger, ensuring that the bottom of the component is kept flat. When the punch returns after completing the draw, this rubber block will assist in ejecting the component.

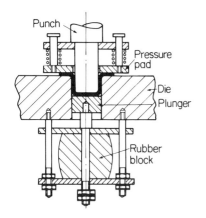

Figure 3.11 Drawing tool

Blank size

The blank for drawing a cylindrical component is a disc. Figure 3.12 shows a cylindrical cup having a diameter d and height h.

Let us now develop a formula so that the blank diameter D may be determined. Ignoring the metal thickness and the radius at the bottom corner

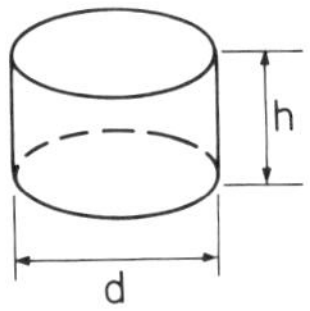

Surface area of cup $=$ surface area of blank

$$\frac{\pi d^2}{4} + \pi dh = \frac{\pi D^2}{4}$$

Figure 3.12 Drawing

Multiplying throughout by 4,

$$\pi d^2 + 4\pi dh = \pi D^2$$

Dividing throughout by π,

$$d^2 + 4dh = D^2$$

Therefore,

Blank diameter $D = \sqrt{(d^2 + 4dh)}$

This formula should only be regarded as a guide, since it does not take into consideration the radius at the bottom corner, and also makes no allowance for stretching that takes place during a drawing operation. The blank size is usually finalised by trial and error, where experience has the last say.

COMBINATION TOOLS Components that are made from relatively thin sheet and drawn into shallow cups may be blanked on the same tool. Figure 3.13 shows a typical combination tool for blanking and cupping.

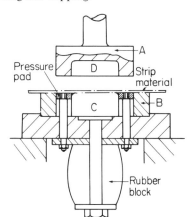

Figure 3.13 Combination tool for blanking and cupping

The punch (*A*) and die (*B*) first blank the material. The blank is then drawn over the punch (*C*) as the punch (*A*) descends containing the cup profile (*D*). To avoid wrinkling the blank is gripped between the bottom of the punch and pressure pad, whose pressure is regulated through the rubber block, which will also assist in ejecting the finished component.

PRESSES Presses may be classified into two main groups: hand operated presses, and power operated presses.

Hand presses The most common type of hand press is the *fly press* shown in Figure 3.14. This is a bench mounted press, used for small presswork operations that require relatively small press forces. It is operated by the handle (*A*), additional energy being provided by the rotary motion of the heavy iron balls, which transfer this energy to the ram via the screw.

Fly presses are often used to prove a new press tool for correct alignment, clearances etc., before being fitted to a power press.

Power presses

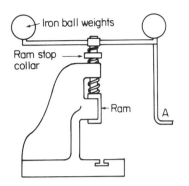

Iron ball weights

Ram stop collar

Ram

A

Figure 3.14 Fly press

The *open fronted press* is perhaps the most common power press, being suitable for both small and medium size work.

The press is operated by a crankshaft and connecting rod mechanism, the connecting rod being fitted to the ram and allowed to move vertically in guides. The connecting rod is adjustable, so that the stroke position can be changed to accommodate different size press tools. Keyed to the end of the crankshaft is a heavy flywheel which provides a reservoir of energy. Each time the press makes one stroke and work is done on the material, energy is given up by the flywheel, and is then restored from the source of power. Some open fronted presses are inclinable (up to about 45°) which permits gravity to be used to dispose of the finished pressings and also allows a chute to be used to feed blanks.

Figure 3.15 shows a typical open fronted inclinable power press.

The *double sided press* is a much heavier machine designed for large work, e.g. body panels for motor vehicles. This form of construction gives greater rigidity but reduces accessibility. The operating mechanism of these presses is almost entirely enclosed. Figure 3.16 shows a typical double sided press.

SAFETY

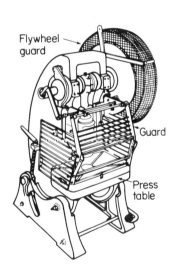

Flywheel guard

Guard

Press table

Figure 3.15 Inclinable open fronted press (by permission of the controller of HMSO)

Both hand and power presses are recognised as being the most dangerous forms of machinery in common use. This is due to the fact that the operator's hands have to enter the space between punch and die to manipulate the work. This has resulted in many accidents, either due to carelessness or accidental starting of the press. A great deal of investigation has been undertaken to establish specific causes of accidents at presses, the following being the most significant.

(1) Press operators obtain a rhythm of working which if disturbed may cause him or her to mistime a press operation when hands are not clear of the trapping area.

(2) Operators tampering with tools due to unexpected jamming of work in die. Under these conditions, the operator should seek the advice of the toolsetter.

(3) Untidy working conditions. Accumulation of waste material and stacking of work on the press table tempts the operator to reach through to back of tools to clear waste and reach work.

(4) Poor layout of press shop, causing jostling of operator, by passing or adjacent workers.

To reduce the risk of such accidents, all presses should be properly guarded by an approved method acceptable to the Factory Inspectorate. The Inspectorate also specify regulations relating to the testing and inspection of presses, and the training of press toolsetters and operating personnel.

PRESS GUARDS

The choice of guard will depend on the press and the nature of work involved but, in general, presses may be guarded by one of the following methods:

(a) The use of enclosed tools
(b) Fixed guards
(c) Interlocked guards
(d) Sweep away devices.

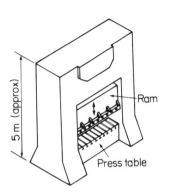

Figure 3.16 Double sided press

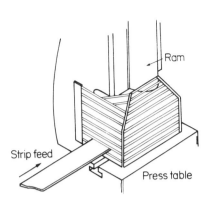

Figure 3.17 Fixed guard

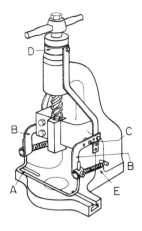

Figure 3.18 Sweep away device
(by permission of the controller
of HMSO)

Enclosed tools   This method is perhaps the simplest way of guarding, but is mainly confined to blanking operations. The tool is designed so that there is insufficient space in the tool area for the trapping of fingers. This is achieved by making the stripper plate sufficiently thick so that the punch is never withdrawn, and also making the feed opening below the stripper small enough, so that fingers cannot reach the punch.

Fixed guards   These guards are simple yet very efficient, since they are a fixture. They consist of a grille construction, so designed to give the operator reasonable vision. These guards are only suitable for blanking operations, or the piercing and forming of blanks, fed to the tools by means other than by hand, e.g. chute feed. Figure 3.17 shows a typical design of fixed guard.

Interlocked guards   These guards consist of a grilled enclosure, the front of which is in the form of a movable gate. This gate is coupled to the press drive mechanism, so that when the gate is open the press cannot be operated, thus permitting safe entry to the tools. There are various ways in which this may be achieved, but basically they all rely on the disengagement of the clutch between flywheel and crankshaft, when the ram reaches the highest point of its stroke. The clutch can only be engaged when the gate is closed. Figure 3.15 shows an interlocked guard system fitted to an inclinable open fronted press.

Sweep away devices   These include safety devices, designed to remove mechanically the operator's hand from the danger area. Figure 3.18 shows such a device fitted to a fly press.
   The sweep rod (A) is secured to the linkage (B) fitted on each side of the press and pivots about (C). The upper end of the linkage is connected to a ring (D) which is fitted between the ram stop and the top of the press frame. As the ram descends the linkage pivots about (C) and causes the rod to sweep from back to front across the lower tool. The spring (E) returns the rod when the ram rises again. The rod can easily be changed to suit the profiles of different tools used.

SUMMARY Presswork includes operations such as blanking, piercing, bending and drawing. Blanking and piercing are both shearing operations. Blanking is performed when the metal removed from the strip is required, while piercing is usually a second operation performed on blanks when holes are required.

Press tools consist of punches and dies which are often given shear to reduce the force supplied by the press. It is important, when designing a press tool for blanking, that a blanking layout is produced to ensure the most economic utilisation of material.

Bending and drawing are both forming operations in which the main material requirement is ductility. Bend allowances are applied to compensate for the deformation that takes place during bending. For vee bends, provision must be made for springback on the punch and die.

Drawing is a process for producing components of dish or cup shape. Some materials for this process may have to be annealed to improve their ductility, since considerable plastic flow takes place. Combination tools are often used to enable both blanking and drawing operations to be carried out on the same tool.

Standard die sets are used to ensure that both punch and die are accurately aligned and also to facilitate fitting to the press.

Presses may be hand or power operated. Due to the dangerous nature of these machines, they must be properly guarded to protect the operator.

QUESTIONS

(1) Explain clearly the difference between:
(a) Blanking   (b) Piercing.

(2) Explain how clearance on the punch and die is applied for blanking and piercing operations.

(3) What is the purpose of shear when applied to press tools?

(4) A 15 mm diameter blank is to be produced from a strip 1 mm thick. If the shear strength of the material is 390 N/mm^2, determine the punch force required.

(5) For a suitable component of your choice, show how this may be blanked from strip to give maximum utilisation of material.

(6) Sketch in good proportion a press tool suitable for blanking and piercing the component shown in Figure 3.19(a).

(7) State what physical property a material should possess for bending.

(8) Explain the term *grain* when applied to sheet metal and state its significance in bending operations.

(9) What is meant by the term *bending allowance*?

(10) Explain what is meant by *springback* when applied to bending operations.

(11) The component shown in Figure 3.19(b) is to be made from steel strip 25 mm wide and having a shear strength of 420 N/mm^2. Determine:
(a) The developed blank length
(b) The punch force required for bending.

(a)
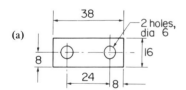

(b)
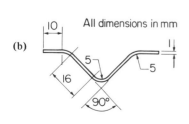

All dimensions in mm

Figure 3.19

(12) Distinguish clearly between bending and drawing.

(13) Using a clear diagram show how a drawing tool is used to produce a cylindrical cup.

(14) Explain the function of the pressure pad on a drawing tool.

(15) A circular metal disc is to be drawn into a cylindrical cup having a diameter of 42 mm and height 20 mm. Ignoring the corner radius at the bottom of the cup, determine the theoretical blank diameter of this disc.

(16) Explain, using a sketch, the function of a standard die set.

(17) Name two types of presses, and state their principle of operation.

(18) Describe three causes of accidents that occur at presses.

(19) Describe using sketches three types of guards found in presses, together with the type of work for which they are best suited.

4 Primary forming processes

INTRODUCTION During its preparation, metal must undergo some process, whereby it is put into a suitable condition so that products can be shaped easily and economically. The origins of some of these processes date back many thousands of years. The principles, however, have remained very much the same, and only the equipment has changed to meet the demands of modern manufacture.

These *primary forming processes*, as they are called, include:

(a)   Rolling
(b)   Casting
(c)   Forging
(d)   Drawing
(e)   Extrusion.

ROLLING This is a very important process since all metals are first prepared by rolling before subsequent shaping operations. Rolling may be carried out with the metal in the hot or cold condition.

Hot rolling When metal is manufactured, it is first cast into large blocks of convenient shape known as *ingots.* An ingot is about 2 m long, tapering slightly along its length, and may be square, rectangular or sometimes of octagonal cross-section. After stripping from the mould, the hot ingot is passed between two heavy rolls so that its cross-section is reduced, while at the same time its length is increased. The rolls are reversible so as to allow the ingot to traverse backwards and forwards through the rolls, the roll opening being slightly decreased between each pass. This initial rolling reduces the ingot into what is known as a *bloom*, which is later cut into convenient lengths for further rolling. The term bloom is used to describe semi-finished hot rolled metal that has an approximate square section of 150 mm or more. Blooms are then re-rolled into *billets* and then *bars*. a Billet refer to semi-finished hot rolled metal having an approximate square section varying between 30 mm and 150 mm square. A bar, on the other hand, refers to a finished rolled section (from billets) and whose length is considerably longer than its width. Billets may be rolled into bars of e.g. square, rectangular, round or hexagonal section, or alternatively of special section, e.g. angle, channel or 'I' section. This is shown in Figure 4.1.

Where the products of the rolling mill are to be in the form of plate sheet and strip the blooms are rolled into slabs, which are rectangular in section, thereby being of better form for re-rolling into flatter shapes. Rolling is carried out in the same fashion as for bars except that parallel rolls are used. Where large quantities of sheet and strip material are required, several pairs of rolls are used. In this way the material is continually passed from one set of rolls to another, so that the material gets progressively thinner until the required thickness is achieved.

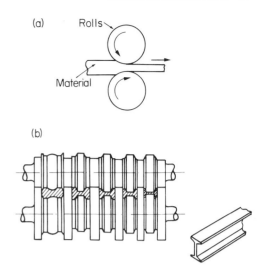

(a)

Rolls

Material

(b)

Figure 4.1 Hot rolling, (a) principle of rolling, (b) rolls showing the stages of hot rolling a steel 'I' section

Steel products produced by hot rolling are always characterised by a reddish blue (oxide) mill scale on the surface of the metal.

Cold rolling This process is often carried out after hot rolling for the following reasons:

(a)   To obtain an improved surface finish.
(b)   To produce products of better dimensional accuracy.
(c)   To improve mechanical properties by combining cold rolling with a suitable heat treatment.

Material for cold rolling is first hot rolled near to finished size and the surface scale removed by *pickling*. Pickling involves immersing the material in a dilute sulphuric acid solution. If this were not performed, then the surface scale would be rolled into the metal, resulting in a poor surface finish. The pressure set up by the rolls is considerably greater compared with hot rolling, since the material is not as ductile. This also means that the reduction in thickness per pass is much less. Cold rolling is carried out using a four high mill as shown in Figure 4.2. The two working rolls are supported by large backing rolls, to prevent distortion of the working rolls due to the high rolling pressures.

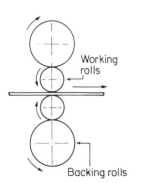

Working rolls

Backing rolls

Figure 4.2 Cold rolling using a four high mill

Cold rolling will always induce a certain amount of work hardening and, if several reductions are required, annealing may be necessary at some stage during the process, to restore the material's ductility. If, after annealing, rolling is regulated then a certain amount of work hardening can bring about an improvement in mechanical properties. This is significant in materials that do not respond to hardening by heat treatment, e.g. certain aluminium alloys.

SAND CASTING *Sand casting* is generally recognised as being one of the oldest methods of shaping metal and dates back to ancient Greek and Roman civilisations.

Sand cast components are produced by pouring molten metal into a mould formed in sand, and whose shape is determined by a pattern. The process consists of three distinct stages:

(1)   Making the pattern
(2)   Making the mould
(3)   Pouring the metal.

Making the pattern   In sand casting the pattern is a wooden facsimile of the component to be cast. The process calls for a high standard of skill on the part of the patternmaker, who will also decide on the best casting technique. This means that the patternmaker must have a sound knowledge of foundry practice. Patterns are usually made from well seasoned yellow pine, since this wood is easy to work and is relatively stable. To allow for contraction when the casting cools, the pattern must be made slightly oversize. For example, the pattern for an iron casting specified as being 100 mm long would have to be 101 mm long, since iron contracts at the rate of 1 mm per 100 mm. To avoid tedious calculations during the making of the pattern, a special rule is used, known as a *contraction rule*, which is so divided to allow automatically for contraction. Different rules would be used for different materials.

Many castings will require machining, which means that an allowance must be provided on the pattern, to provide extra metal at the relevant surfaces. To facilitate removal from the mould, patterns are often split along a parting line the position of which will be governed by the shape and complexity of the casting. The two half patterns are located together by dowels. As a further aid to removal, patterns should also have slightly tapered sides (2–3°), known as draught angles.

Making the mould   Let us consider the casting of the component shown in Figure 4.3. A half pattern is first laid face down on a moulding board and a moulding box known as a *drag* is positioned around it. The pattern surface is then covered with *facing sand*, which is fine unused sand, to give the casting a good surface finish. The remainder of the box is filled and rammed up with *backing sand*, sometimes known as *greensand*, not because of its colour, but because it is slightly moist. With the excess sand skimmed off the half mould (drag) is at the stage shown in Figure 4.4(a). The drag is now inverted and the top half of the pattern located, and another moulding box known as the *cope* is located on to the drag. *Parting dust* is then sprinkled over the mould face so that the half moulds are easily separated when the pattern is removed. A *runner* (pouring hole) and a *riser* must be provided. The riser will indicate when the mould cavity is full and also exhaust the air and any hot gases produced. The runner and riser are formed by positioning tapered wooden plugs in the cope, and later removed when the cope has been rammed up with sand, in a similar fashion to the drag. This stage is indicated in Figure 4.4(b).

The cope is now separated from the drag and the two half patterns carefully removed. This is achieved by screwing into the pattern faces a

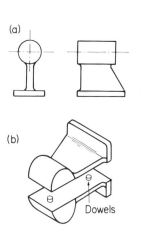

Figure 4.3 Pattern for sand casting, (a) component to be cast, (b) split pattern

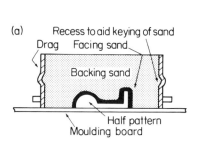
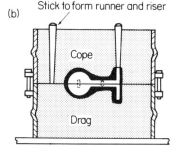

Figure 4.4 Sand casting mould

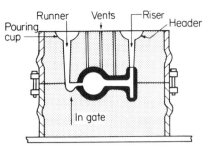

Figure 4.5 Mould ready for pouring

rod known as a *rapping bar*. The moulder then taps this in several directions so that the pattern is loosened, thus permitting a clean lift without disturbing any surrounding sand. A channel known as an *in gate* is now cut, to connect the runner with the mould cavity. To assist the escape of gases which are formed when the molten metal comes into contact with the damp sand, venting will be required. This is done by inserting a wire in the sand to form small holes at several places, but stopping short by about 4 mm from the mould cavity. The mould cavity is then dusted with a graphite compound called *plumbago*, which assists the facing sand in providing a good finish to the casting. The cope and drag are now reassembled. A pouring cup and header are placed over the runner and riser respectively. These may be made separately or cut away from the cope sand. The mould is now ready for pouring as shown in Figure 4.5.

Pouring the metal

The quality of a casting can be greatly influenced by the way the metal is poured. This should be done as rapidly as possible, but at a steady rate without overflowing the pouring cup and continued until the header becomes full. The header will also maintain a pressure in the mould, ensuring that it is completely full.

Fettling

When the casting has solidified and the sand is broken away, the casting will have a number of unwanted projections in the form of runner, riser and in gate as shown in Figure 4.6.

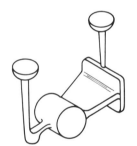

Figure 4.6 Casting when removed from the mould

The removal of these projections and the general cleaning up of the casting is known as *fettling*, which is performed using chisels, files and grinding wheels. Shot blasting is often used whereby steel shot or coarse particles of sand are projected at the casting within a stream of compressed air to remove moulding sand adhering to the casting.

Casting hollow sections

The majority of castings are not completely solid but may include a number of holes or hollow sections. Where a large hole, for example, is to be machined in a casting, this may be cast by making it smaller than finished size, to allow for machining. This not only saves metal, but will also decrease the machining time and hence lower the manufacturing cost.

To cast a hole, a core will be required which represents the pattern for the hole and is made from sand mixed with a suitable resin binder. The core is formed in a wooden core box, made in two halves and located together using dowels. Using simple clamps to hold the two halves of the core box together, the box is filled with sand and rammed up. After ramming the box is then split and the moulded core baked in an oven to increase its strength, so that it is able to withstand the in-rush of molten metal. The pattern required for the casting will require two additional

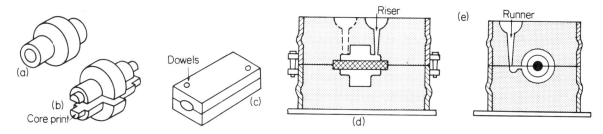

Figure 4.7 Casting a hole, (a) component, (b) split pattern, (c) core box, (d) side view of mould, (e) end view of mould

features known as *core prints*. These will leave impressions in the sand when the pattern is removed so that the core can be located. It is important that the core is positioned carefully since a misplaced core will result in the casting having an uneven thickness. Figure 4.7 shows an example of the preparation of a core and its location within the mould.

Defects in castings   There are a number of defects that are sometimes found in castings.

Scabs. These are formed when a portion of sand breaks away from the mould wall and results in an irregular protrusion being left on the surface of the casting.

Blow holes These may be attributed to a number of causes, but are mainly due to insufficient venting, pouring the metal at too low a temperature or the metal having an incorrect composition.

Cracks Attributed mainly to faulty design, i.e. uneven sections and the presence of sharp corners instead of generous radii. Too high a pouring temperature may also lead to cracking.

Characteristics of castings   When molten metal is poured into the moulds, the metal first making contact with the mould wall will solidify more rapidly than the rest. This is due to the sand being at a much lower temperature than the metal; it therefore has a chilling effect, resulting in a fine grain size at the outer layers of the casting. As the casting continues to solidify, which will now be much slower, grains will form in the direction of cooling, and will always be associated with a coarser grain size. This type of structure does

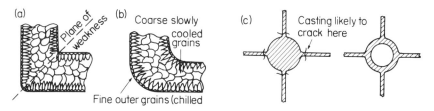

Figure 4.8 Casting characteristics, (a) sharp corners in castings are liable to cracking along plane of weakness, (b) effect of radius at corner to give improved strength, (c) improvement of design: large mass removed by casting a hole

not always possess desirable properties. While this structure cannot be controlled during the manufacture of the casting, it may however be improved by a suitable heat treatment, such as annealing or normalising, depending on the properties required. Figure 4.8 shows a typical cast structure and its relation to strength at corner sections.

The chilled outer layers will also contain small particles of sand which together form a hard skin. When machining, this skin should be penetrated during the first cut, since failure to do so would lead to rapid tool wear.

When a casting cools, it will contract by varying amounts due to parts having different masses. This will lead to internal stresses, which if excessive may cause cracking, as in Figure 4.8(c). Although the pattern-maker will try to minimise these stresses when analysing the best casting techniques to adopt, they cannot be completely eliminated at the casting stage. It is important that some castings are stress free, e.g. surface plates and machine tool structures. These are often rough machined first, followed by *seasoning*. Seasoning involves leaving the casting out in the open to weather, so that internal stresses are relieved naturally. On small castings, seasoning may be carried out artificially by tumbling. The castings are loaded into a tumbling barrel whereby the light blows that they receive when rotating help to relieve some of the internal stresses.

In general most metals may be sand cast; the main physical property required is fluidity, which is essential if good cast impressions are to be made. Fluidity may be improved by adding certain elements to the cast metal. Silicon is a notable example, which is included in cast iron and some aluminium alloys.

FORGING

Like casting, forging also has ancient origins that have become a well established technique and adapted to suit the requirements of modern manufacture. Forging basically involves heating a bloom billet or bar so that its plasticity and malleability are increased and then shaping it by squeezing, bending or hammering.

Hand forging

This is also known as *smith forging* and is used to describe the shaping of a range of small work using relatively simple tools, including hand hammers. Power hammers are sometimes used. The quality of work produced is solely dependent on the skill of the smith, who sometimes works with an assistant who acts as the striker when large hammers are required. The

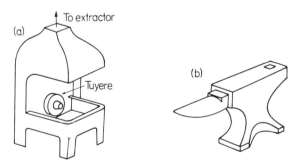

Figure 4.9 Blacksmith's equipment, (a) hearth, (b) anvil

two fundamental pieces of equipment used are the *hearth* and *anvil* as shown in Figure 4.9.

The hearth is the means by which the work is heated and is coke fired. The temperature is increased using a water cooled *tuyere*, which supplies air under pressure from a separate blower. The anvil is used to provide support for the work during forging. The main body consists of a steel casting with a hardened top surface. The *beak* is left soft and is so shaped to permit the bending of different radii. Opposite the beak is a square hole to locate various blacksmith's tools.

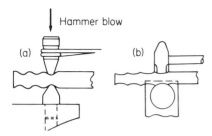

Figure 4.10 Drawing down, (a) using fullers, (b) on beak of anvil

The following represent some common hand forging operations.

Drawing down   The purpose of drawing down is to reduce the cross-sectional area of the metal with a corresponding increase in length. This is achieved by using tools known as *fullers* or hammering the metal on the beak of the anvil, as shown in Figure 4.10.

Setting down   This operation is similar to drawing down except that the metal is made to spread, so that both length and width are increased. Setting down, shown in Figure 4.11, is often performed after drawing down, to remove the wavy surface caused by the latter.

Upsetting   This operation causes the metal to increase in cross-sectional area with a corresponding decrease in length. Only the portion to be upset is heated; the material is then stood end on, to be struck by a hammer, as in Figure 4.12.

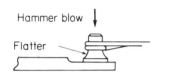

Figure 4.11 Setting down

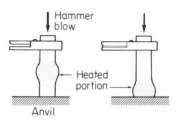

Figure 4.12 Upsetting

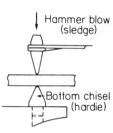

Figure 4.13 Cutting

Cutting   This process, shown in Figure 4.13, is used for removing pieces of metal from a bar using chisels. Cutting is usually started on one side of the work and completed from the other side.

Swaging   This process is used when the work has to be reduced and finished to a specified section, e.g. round, hexagonal. Swages come in pairs, as shown in Figure 4.14, either separate or held together by a spring steel strip.

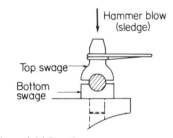

Figure 4.14 Swaging

Forging presses   On very large components that require forging, hammer blows will have very little effect since they will only be felt at the outer layers of the metal. To overcome this *forging presses* are used, which are very large machines, operated hydraulically, and designed to apply a steady squeeze to the metal. The forces used are considerable, 80 MN being a typical figure, although some presses are in existence rated at 175 MN or more. Very simple tools are used, vee tools being the commonest and used for forging round work and where large reductions in sections are required. Large cylinders and rings, e.g. 5 m in diameter, may be forged. These are first prepared by hot piercing or machining a small hole in the ingot or bloom and then forged over a mandrel. This operation is known as *becking*. The movement of the press ram together with the mechanical manipulation of the forging is directed by a *forge master*, who conveys his instructions using hand signals.

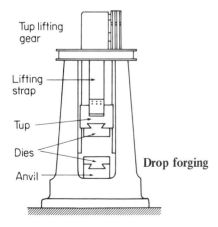

Figure 4.15 Drop hammer

Drop forging   Although hand forging is a useful shaping process for small components, it is however not suitable for the manufacture of a large number of identical forgings. For large scale production forging, *drop forging* is employed. The process consists of placing a hot metal billet between two dies, whose impression corresponds to the component to be forged. The lower die is fixed and the upper die lifted and then allowed to drop on to the billet placed on the lower die. The component is thus formed to shape by a succession of blows, about three or four, depending on the complexity of the component. The machine used for drop forging is known as a *drop hammer* or *drop stamp.* The lower die is fixed to the anvil, while the upper die is fixed to the tup and is allowed to fall between a pair of guides attached to the machine frame. Figure 4.15 shows a typical drop hammer.

Drop forging dies

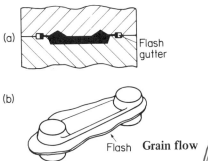

Due to the severe stresses imposed on dies, these are made from special die steels which contain small amounts of chromium and nickel. These alloying elements give increased die life, together with the required properties after heat treatment. Die impressions must be provided with a draught, usually about 7°, so that the component is removed without sticking. In addition, it is essential that a *flash gutter* is provided as shown in Figure 4.16. This will accommodate the excess metal when the die faces come together, indicated by a characteristic ringing sound. This excess metal will leave a flash or fin around the forging, which is later removed on a shearing press.

Grain flow

A forged component usually has better mechanical properties, compared with the same component that has been cast or machined from solid. This is attributed to the characteristic *grain flow* or *fibre* which is the result of the original grain structure being deformed, thus producing a new structure, which follows the general shape of the component. The properties of this structure are similar to the grain found in a piece of wood, i.e. the maximum strength is along the grain rather than across it. This means that the directional characteristic of the grain may be used to maximum advantage, especially on highly stressed components.

Figure 4.16 Drop forging, (a) die, (b) drop forged connecting rod showing flash

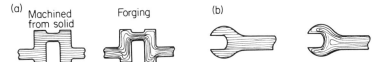

Figure 4.17 Grain flow, (a) crankshaft, (b) spanner

Figure 4.17 shows some examples of forged components and also how the grain flow compares with that of the same component machined from solid.

COLD DRAWING

Wire and much of the bar material used in manufacture is finished to size by *cold drawing.* The material used for this process must be extremely ductile and is first prepared by hot rolling. The material after rolling will be covered in mill scale, which must be removed by pickling before drawing commences.

Drawing involves pulling the bar through a conically shaped die, so that its diameter is reduced and its length increased as shown in Figure 4.18.

In many cases, to achieve the finished size, the bar will have to be passed through a succession of dies, each getting progressively smaller. Due to the extreme pressures set up between the metal and die, it is usual to lubricate the area of contact. This is done by passing the undrawn bar through a container filled with lubricant before it enters the die. Typical lubricants used are grease and certain soap solutions. Steel bar that has been cold drawn is always characterised by a bright surface finish and is often referred to as *bright drawn steel*.

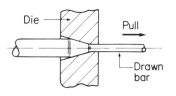

Figure 4.18 Principle of drawing

In the case of wire drawing where considerable reduction in diameter is required, work hardening will be present. This means that the drawn material will have to be annealed at various stages during the process.

Typical materials suitable for cold drawing include low carbon steel (mild steel), copper, brass and aluminium.

EXTRUSION

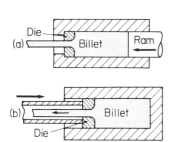

Figure 4.19 Extrusion, (a) direct, (b) indirect

This process is used to produce long lengths of continuous section (often complex) in non-ferrous metals such as brass, copper and aluminium alloys. Non-ferrous metals are suitable due to their low compressive strength and high plasticity at elevated temperatures. The process consists of placing a heated billet into a pre-heated container and forcing it through a die, whose profile is the same as the section required. The force is supplied by a hydraulically operated ram.

Extrusion is classified as *direct* or *indirect*, as shown in Figure 4.19. The indirect method is generally preferred since the ram force is lower due to the absence of frictional forces, which are present in the direct method, caused by the billet moving relative to the container walls. The direct method also has the disadvantage that some surface scale is drawn into the centre of the last part of the billet. This means that not all the metal can be extruded, if this is to be avoided.

Impact extrusion

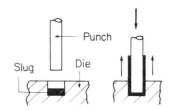

Figure 4.20 Impact extrusion

Unlike the extrusion process already discussed, this is performed cold on soft metals such as aluminium and lead.

In this process a piece of preformed metal known as a *slug* is placed in a die. This slug is then struck by a punch moving at high speed (about 25 m/s), which causes the material to become plastic and flow back up the punch. The thickness of the extruded section will be dependent on the clearance between the punch and die.

Typical products produced by this process include cigar tubes, film cans and protective cases for electrical components. Figure 4.20 shows the principle of impact extrusion.

SAFETY

Primary forming processes often require the manipulation of hot metal and in many cases under the influence of large forces. In this respect certain safety precautions should be observed.

Clothing

As with all workshop activity protective clothing is essential, especially with regard to footwear, and approved safety shoes must be worn. When pouring molten metal, safety glasses with tinted lenses should also be worn.

Hot castings and forgings

When castings are broken away from their moulds and forgings removed from dies etc., provision should be made for them to cool down. Metal baskets labelled HOT METAL are sufficient and should be positioned so that they can do no harm.

Handling molten metal

When pouring molten metal into moulds, this should be done as close to the furnace as possible, so that loaded crucibles are not moved any distance. Lifting equipment, e.g. tongs and cradles, must fit pouring crucibles properly and provide adequate support. When pouring large castings the molten metal may have a tendency to lift the cope sand off the drag. This should be anticipated in advance and weights placed on top of the cope.

Ventilation Although all workshops should be properly ventilated, extra precaution should be observed in foundries for the following reasons:

(1) Due to the nature of the materials used in foundries, i.e. sand, the atmosphere is prone to dust contamination.

(2) Melting furnaces and moulds during pouring give off fumes which must be efficiently extracted.

SUMMARY Primary forming processes are performed to shape metal so that further manufacture can be carried out both easily and economically. Metal is first cast into ingots and then reduced in size by hot rolling to form blooms, slabs and billets. Hot rolling is also used to produce continuous lengths of steel section, e.g. angle, 'T' and channel. Where dimensional accuracy, surface finish and improved physical properties are important, hot rolled metal is often finished by cold rolling.

Sand casting is used to shape components by filling a sand mould with molten metal. The mould is first prepared using a suitable pattern. Where hollow sections are required, cores are made using a core box, and located in the mould using prints left by the pattern. Sand castings are completed by fettling and may require seasoning to relieve internal stresses set up during solidification.

Forging consists of hammering or squeezing hot metal to shape. Hand or smith forging employs simple tools, and may be used to forge small components that are not required in large quantities. Where quantities are large then drop forging is used, whereby the component is shaped between two dies which are subjected to a series of blows. Forging produces a characteristic grain structure, which often has better physical properties compared with a cast structure, or material machined from solid.

Rod, bar and wire are usually finished by cold drawing. This involves pulling the metal through a series of reducing dies until the required size is obtained.

Hot extruded products are produced by pushing a hot metal billet through a die, whose profile corresponds to the section required. This process is only used for non-ferrous metals such as copper, brass and aluminium alloys. Impact extrusion is carried out cold, and is used on soft materials, e.g. aluminium and lead, to form thin walled hollow components.

QUESTIONS

(1) Compare the processes of hot rolling and cold rolling with regard to surface finish, dimensional accuracy and physical properties of the rolled section.

(2) Explain using a diagram what is meant by a four high mill.

(3) With reference to casting, explain what is meant by the following terms:
(a) Pattern   (b) Cope and drag
(c) Core   (d) Fettling.

(4) Explain why a sand mould must be vented prior to pouring.

(5) Name three defects that may be present in a casting and state their possible causes.

c

(6)   When a hole or hollow section is to be produced in casting, explain what provision must be made on the pattern.

(7)   Explain what is meant by seasoning when applied to castings.

(8)   With the aid of sketches show what is meant by the following forging operations:
(a) Drawing down
(b) Upsetting
(c) Swaging.

(9)   Explain, using a sketch, the principle of drop forging and state the type of work for which this process is suitable.

(10)   Describe with the aid of a sketch, what is meant by a flash gutter in a drop forging die.

(11)   Name two material properties that are required if a component is to be forged successfully.

(12)   For a component of your choice, give two reasons why a forging is more beneficial than machining from solid.

(13)   Describe using a sketch the process of cold drawing. State what physical property is required of the material for this process.

(14)   Using a sketch show the difference between direct and indirect extrusion. Give two reasons why the indirect method is to be preferred.

(15)   Describe, using a sketch, impact extrusion and name two examples of components produced by this process.

force is directed through the tool tip and into the body of the tool, thus providing sufficient support.

Although the shear plane has been increased the cutting force may be kept to a minimum by using high cutting speeds, which under these conditions promotes plastic flow of the work material, thus decreasing the force on the tool. Since the cutting force has been reduced the power requirement has been increased due to the high speeds necessary, which demands more power than that saved by the reduced cutting force.

If negative rake is to be successful then the following conditions should be observed:

(1) Adequate power available to machine
(2) Adequate range of cutting speeds
(3) Maximum rigidity for both cutting tool and work
(4) Good surface finish on top face of tool
(5) Cutting should be continuous.

Although negative rake was introduced due to the brittle nature of cemented carbides and ceramics, it is now a well established technique when using throw-away tips. This is because a throw-away tip used for negative rake cutting can be given twice as many cutting edges, compared with a similar tip used for positive rake cutting. This will be appreciated more readily in the next section on tipped tools.

An interesting feature when using negative rake is that the chip often leaves the work red hot, but the tool and workpiece are relatively cool, and a good surface finish is produced on the work.

SINGLE POINT TIPPED TOOLS

Modern machining methods employ the use of carbide and ceramic tooling, and although these tool materials have efficient cutting properties they are, as already stated, brittle and hence require a special type of toolholder.

Cemented carbide tipped tools

Nowadays these tools fall into two main groups: *brazed tipped tools*, and *indexable* or *throw-away tips*.

BRAZED TIPPED TOOLS
These tools are used extensively in production, and are commonly used in shaping, planing and turning operations. Many shapes and sizes are available with different tool angles to suit different work materials.

They are produced by machining a recess into the tip of a medium carbon steel shank and then brazing in the carbide insert. The brazing material is in the form of a shim, so that when it melts it flows by capillary action, between the shank and the insert. So that heating is uniform and the risk of scaling eliminated a high frequency induction heating coil is often used, as shown in Figure 5.18(b).

INDEXABLE OR THROW-AWAY TIPS
These tools are rapidly becoming popular, and offer the following advantages compared with brazed tipped tools.

(1) The design of the tool holder allows a number of cutting edges to be used on the same tool, simply indexing the tip to the next position when one edge has become dulled. By turning over a three or four sided tip, six or eight cutting edges respectively may be provided, which can only be achieved if negative rake is used.

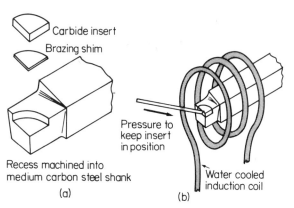

Figure 5.18 Preparation of carbide tipped tool, (a) components, (b) tool is heated by eddy currents induced as a result of magnetic field produced by high frequency current in coil

(2)   After all edges have been used, the tip is thrown away and replaced by another. Taking into account the initial cost of the tip, this is more economical since the non-productive time/cost of re-grinding a brazed tipped tool would be greater — an important consideration in batch and volume production.

(3)   Chip breakers may be built into the toolholder, which may be adjusted to suit metal cutting conditions.

Due to the difficulty of brazing ceramics into steel shanks, they are always used in the form of throw-away tips. This difficulty now becomes insignificant due to the advantages offered by throw-away tip tooling.

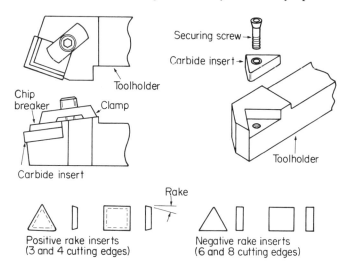

Figure 5.19 Throw-away tip toolholders and inserts

Figure 5.19 shows the construction of a typical throw-away tip toolholder and the type of inserts available.

QUICK CHANGE AND PRESET TOOLING   When tools are used to manufacture large quantities, it is important to keep re-grinding and tool setting time to a minimum, so that the process remains economic. These times may be minimised by using tools that have been *preset* and used in conjunction with *quick change* toolholders, so that a duplicate tool can be substituted when it is necessary to re-grind an existing tool.

The design of quick change toolholders varies, but should be such that they can be easily fitted, and also give precise repeatability of

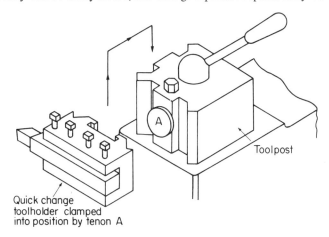

Figure 5.20 Quick change tooling

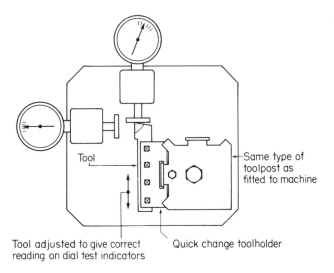

Figure 5.21 Preset tooling fixture

position. This is important if the dimensions of the machined work are governed by stops to control the movement of the machine slides. Figure 5.20 shows a typical arrangement for a quick change toolholder.

When duplicate tools are used, then these must be set using a preset tooling fixture, so as to ensure repeatability when the tool is fitted to the machine. Figure 5.21 shows a typical fixture used in conjunction with dial test indicators.

CUTTING FLUIDS   During any metal cutting operation, friction and heat are generated, both of which may have adverse effects on the workpiece and cutting tool. Because of this it is desirable in many machining operations to introduce a *cutting fluid* to the area of cut, which will have both cooling and lubricating properties.

The exact nature of the cutting fluid chosen will depend upon the machining process and the material to be machined, but in general a cutting fluid should ideally possess the following properties:

(1)   Cool work and cutting tool
(2)   Reduce friction between work and tool
(3)   Improve surface finish
(4)   Increase tool life
(5)   Reduce cutting force and power consumption.

There are two basic types of cutting fluids: *soluble oils* (coolants) and *straight oils*.

Soluble oils   These are emulsions of oil and water used chiefly as coolants with limited lubricating properties; they find application in general machining for both ferrous and non-ferrous materials. The ratio of the oil/water mix varies depending on the application, but is usually between 5 parts water to 1 part oil, and about 50 parts water to 1 part oil. When mixing soluble oils the manufacturer's instructions should be adhered to and in all cases must be carried out prior to supplying the machine. The correct mixing procedure can only be obtained by adding oil to the full amount of water while continuously stirring. If water is added to the oil then the correct emulsion will not be achieved.

Straight oils   These oils are ready for use as supplied and are used where lubrication between work and tool is more important than cooling properties, although straight oils do offer a limited amount of cooling.

Straight oils are blended from two main groups of oils: *mineral oils* and *fatty oils* (oils of animal or vegetable origin).

These oils are seldom used on their own for the following reasons. Mineral oils have good lubricating properties, but tend to break down under heavy pressure. Fatty oils are able to withstand heavy pressures but have the disadvantage that they may chemically attack the work-piece material and become rancid after prolonged use. Straight oils are therefore usually blended by mixing a small quantity of fatty oil with a mineral oil, so that the advantages of each type are combined in the one oil. The mixing proportions will be dependent upon cutting conditions, material to be machined, etc.

Blended oils with EP additives   Straight oils used under arduous conditions may break down, i.e. the oil is squeezed out from between the chip and tool face, thereby destroying its lubrication properties. To overcome this problem *extreme pressure (EP) additives* are often included during the preparation of straight oils.

The two main EP additives are sulphur and chlorine. Although sulphur is a useful additive it does have the disadvantage that it will attack copper rich metals, e.g. brass, forming a stain on the newly machined surface. The extent of this staining depends on the amount of free sulphur present and the time that the oil is in contact with the work. In this respect the sulphur content must be carefully controlled.

Chlorine is not so widely used as sulphur since it will chemically attack steel, although it produces no harmful effects on copper rich alloys. Some oils contain sulphur and chlorine, often referred to as *sulpho-chlorinated EP oils*, which are suitable for severe machining operations on high tensile, stainless and heat resisting steels.

Application of cutting fluids

Although the correct cutting fluid may have been chosen it is important that it is applied efficiently between the cutting tool and the work in order to obtain the maximum desired cooling or lubricating properties. As a general rule cutting fluids should be applied in sufficient quantity at low pressure. If high pressure is used this is likely to cause splashing and will be wasteful and hence uneconomic. In some cases high pressure may be advantageous especially when machining deep bores, whereby the high pressure will assist in the removal of swarf.

Most modern machine tools have a built in recirculatory cutting fluid supply system which allows the fluid to be pumped from a nozzle in the form of an adjustable pipe. Whilst this type of nozzle is satisfactory for most machining operations, it may be beneficial in some cases to use a flattened nozzle so that the cutting fluid is fanned out over a large area. This is particularly significant in machines employing several tools that are used to cut simultaneously. Flattened nozzles are often found on grind ing machines where the cutting fluid is required to cover the whole width of the grinding wheel.

On machines using a pumped supply it is important that the cutting fluid is properly filtered, the filters being cleaned at regular intervals. Much blame is often attributed to supply systems simply because the filters become blocked. Furthermore if the fluid supply is to remain effective then the tank holding the fluid (sump) should never fall below a level whereby the pump ceases to be efficient.

Cutting fluids and health

Providing care is taken with regard to personal hygiene and simple precautions are observed, the handling of cutting fluids does not present any real problem. However, continual improper use, by allowing cutting fluids to come into contact with the skin, could lead to industrial dermatitis. This is caused by skin irritants becoming lodged in the pores of the skin causing irritation and inflammation of the hands and forearms. Such irritants may be present in cutting fluids in the form of small metallic particles or fine abrasive particles such as those produced by grinding operations.

When working with cutting fluids the following precautions should be observed:

(1)   A suitable barrier cream should be used before work is started and any cuts to the hands suitably protected.

(2)   After work is complete, hands and forearms should be thoroughly washed using hot water and soap. Solvents such as paraffin and white spirit should never be used, as these themselves may act as skin irritants.

(3)   Oil soaked clothing should never be worn, nor should oily rags be put into pockets which could soak through to the skin.

SUMMARY

During metal cutting the type of chip produced will depend on the material being cut and various cutting conditions. In general the chip will either be continuous, discontinuous or continuous with a built up edge. The latter should be avoided. Three forces act at the tool tip during cutting and are of significant interest from the point of view of power consumption and tool life. These forces are dependent on cutting speed, feed, tool geometry and depth of cut. Tool dynamometers are used to measure these forces.

Various types of materials are used for cutting tools which include high speed steel, tungsten carbide, ceramics and diamonds. High speed steel is the most common and is available in different types. Tungsten carbide may be used as a brazed insert, or together with a ceramic tip used in the form of an indexable throw-away tip. Due to the brittle nature of tungsten carbide and ceramics they are often given negative rake. Diamond cutting tools are used for machining non-ferrous metals at high speed and require a special machine. On production machining quick change and preset tools are frequently used.

Cutting fluids are used primarily to increase tool life, improve surface finish and reduce power consumption. Soluble oils (coolants) are used to cool both tool and work while straight oils are used where lubrication is more important than cooling. These fluids should be applied efficiently using the correct design of nozzle. Care should be observed when handling cutting fluids with regard to health.

QUESTIONS

(1) State what effect speed, depth of cut and tool geometry have on cutting force and power consumption.

(2) List the factors that affect the power consumption during a metal cutting operation.

(3) During a test using a lathe tool dynamometer the following results were recorded:

Spindle speed 500 rev/min   Feed 0.4 mm/rev
Blank diameter of work 60 mm   Cutting force 1650 N
Depth of cut 5 mm   Feed force 475 N

Determine the power consumed at the tool point.

(4) With reference to cutting tool materials, what is meant by hot hardness?

(5) State the approximate composition of a typical high speed steel.

(6) State, giving suitable reasons, the main area of application of diamond cutting tools.

(7) Explain why cemented carbide and ceramic tipped tools are usually given negative rake.

(8) What are the essential requirements if negative rake is to be used successfully?

(9) What is meant by the term built up edge?

(10) Explain what factors should be taken into account before deciding on a metal removal rate.

(11) The following information applies to the machining of a certain steel bar in a centre lathe:

Cutting speed 30 m/min
Depth of cut 4 mm
Feed 0.2 mm/rev

Determine the metal removal rate in mm^3/min.

(12) State the three forces that act at the tip of a single point cutting tool.

(13) State the advantages offered by throw-away tip tooling.

(14)   With the aid of a suitable sketch, describe a typical design for a throw-away tip toolholder.

(15)   An HSS tool is found to have an average life of 1 hour between re-grinds when used in a lathe for roughing cuts on steel bar at 0.5 m/s. Given that $n = 0.12$ for roughing and 0.10 for finishing, calculate the probable life for the tool between re-grinds when used for finishing cuts on the same steel bar.

6 Machine tools

INTRODUCTION The primary function of a machine tool is to remove metal under power, so that the component is finished to the correct shape and size.

The most important of all machine tools is the centre lathe, the origins of which have been found in the records of Leonardo da Vinci (1452–1519), who recognised the need for such machines to enable him to construct his own inventions. It was not until 1810 when Henry Maudsley, generally regarded as the father of the modern machine tool, constructed the first screw cutting lathe. This then enables other machine tools to be made and by the mid 1800s the lathe became a well established production tool, due mainly to the industrial changes taking place in Britain at that time.

The lathe formed the basis from which other machine tools have developed, e.g. milling machines and shaping machines. This development is by no means complete, since the requirements of modern technology demand machines having higher output capabilities together with improved efficiency.

The purpose of this chapter is not to treat machining processes in depth but to consider the range of machine tools used in manufacture together with their principles of application.

PRODUCTION OF MACHINED SURFACES Components produced by machine tools have surfaces which are either flat or curved, which may be produced by a variety of machining operations using many different types of cutters. These operations can however be grouped into two main types, known as *generating* and *forming*.

Generating A surface produced by generating is one that has been shaped by the relative movement of the work and cutting tool. A generated surface is

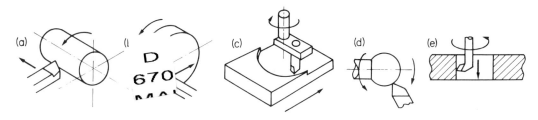

Figure 6.1 Generating operations, (a) plain cylindrical surface, (b) flat surface, (c) flat surface, (d) sphere, (e) parallel bore

completely independent of the tool shape. Figure 6.1 shows some examples of this principle.

Forming A surface produced by forming is one whose profile is solely dependent upon the shape of the cutter used. Figure 6.2 shows some common examples. It will be noticed that these formed shapes are also partially dependent upon generating principles, i.e. there must be relative movement between the work and cutting tool.

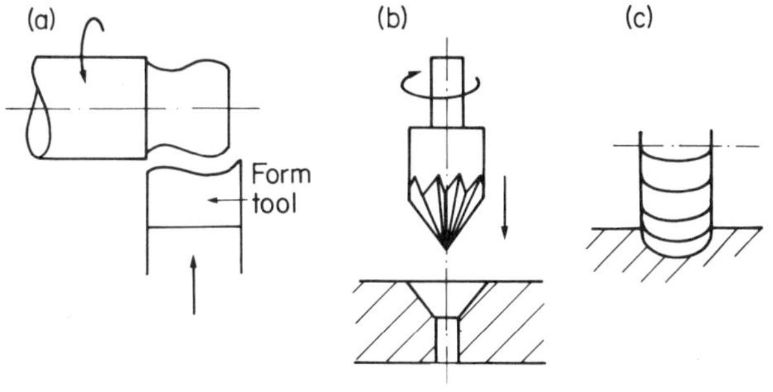

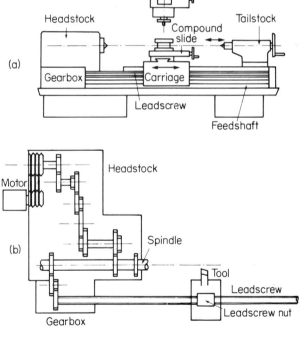

Figure 6.2 Forming operations, (a) shaping, (b) countersinking, (c) milling a concave slot

THE CENTRE LATHE   This machine is designed to produce components of cylindrical shape which may be of external or internal form. This is achieved by the work being suitably held and rotated relative to a single point tool, which moves parallel or at right angles to the longitudinal axis of the machine. Tapered components can also be produced by arranging the tool to move at an angle to this axis. Figure 6.3 shows the basic layout of a centre lathe together with its transmission.

Figure 6.3 Centre lathe, (a) basic layout, (b) simplified layout of transmission from input to cutting tool

The *bed* forms the main structure of the machine to which all the other parts are mounted. It is made from a good quality alloy cast iron and ribbed along its length to provide sufficient stiffness so that it is able to withstand the stresses imposed upon it. The top surface is planed to provide guideways for the carriage and tailstock.

The *headstock* consists of a box type casting and houses the driving arrangements from the motor to the spindle. The spindle is hollow so that bar may be passed through it and also provides facilities for attaching work holding devices. The headstock contains the drive mechanism from the spindle to the feedshaft and leadscrew via a gearbox, thus permitting power feeds to be used and screw cutting to be performed

The *tailstock* is at the other end of the bed and opposite the headstock Its function is to support a centre when turning between centres and also tools for drilling and reaming. It is important that the axis of the tailstock is in line with the spindle axis.

The *carriage* is mounted on the bed and provides the longitudinal movement of the cutting tool parallel to the axis of the machine. The carriage consists of two parts, the front known as the *apron* and the top known as the *saddle.* Mounted on the saddle is the *cross slide* and *compound slide.* The compound slide is only used for the machining of short tapers.

Work holding devices include chucks of various types, e.g. the three jaw self centring chuck for holding round bar, and the four jaw independent chuck for holding irregular shaped work. *Collet chucks* are used for holding accurate round bar of standard size. So that the accuracy of these collets is maintained they should only be used with the size of bar for which they were designed. In addition to chucks, work may also be held on a *faceplate*, which is suitable for large work that cannot be easily

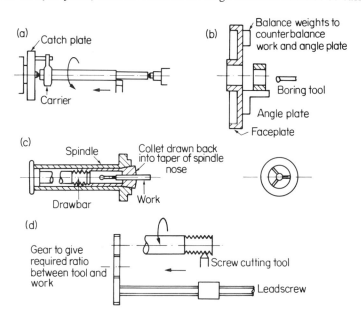

Figure 6.4 Typical work holding devices and turning operations, (a) plain cylindrical turning between centres using catch plate and carrier, (b) boring using a faceplate, (c) collet chuck, (d) screw cutting

held in a chuck. Long work is usually turned between centres and driven by a *catch plate* and *carrier.*

Figure 6.4 shows some common turning operations together with various work holding devices.

Screw cutting   Lathes equipped with *leadscrews* permit screw cutting to be carried out. The principle of any screw cutting operation is to arrange for the tool to move by a controlled amount (parallel to the longitudinal axis of the machine) relative to the rotation of the work. This is achieved by using suitable gearing between the machine spindle and leadscrew. Modern lathes are provided with a gearbox so that the required gear ratios are selected easily for a given thread pitch. Lathes not equipped with a gearbox are provided with change wheels to enable the required gearing to be calculated, which is then set up by the operator.

The principle of screw cutting is illustrated in Figure 6.4(d).

Centre lathes are versatile machines which are set and operated by a skilled operator.

The *capacity* of a lathe is specified as the distance between centres and its centre height from the bed. Sizes may range from about 300 mm between centres by 50 mm centre height to 10 000 mm X 1500 mm.

The use of centre lathes is mainly confined to prototype and small batch manufacture.

CAPSTAN AND TURRET LATHES

These machines are a development of the centre lathe, so that turned parts can be produced rapidly using pre-set tooling. The general design of these machines is similar to the centre lathe, except that the tailstock is replaced by a *turret*. This usually has six tool stations, which can be indexed to present different tools to the work. In addition, the cross slide has five tool stations, a front four way tool post and a single rear tool post. The movement of these tools is controlled by adjustable stops. Figure 6.5 shows the general design for both capstan and turret lathes.

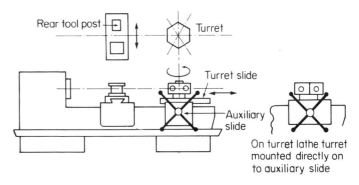

Figure 6.5 Capstan and turret lathes

On the capstan lathe the turret is carried on a separate slide, which when moved back (operated by a star wheel) will index the turret to the next tool station. On the turret lathe, the turret is mounted directly on to the saddle. This creates a far more rigid construction, thus making turret lathes more suitable for heavy work, e.g. the machining of castings and forgings.

The headstock transmission on capstan and turret lathes differs slightly compared with that of the centre lathe, to permit rapid speed changing even while the tools are cutting. This is often achieved by using a constant mesh all geared headstock, where gear changing is performed by operating multi-plate *friction clutches*. This principle is shown in Figure 6.6.

Capstan and turret lathes are usually set by skilled toolsetters and the machines operated by semi-skilled operators.

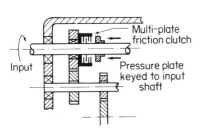

Figure 6.6 Principle of constant mesh headstock. When pressure is applied in the direction shown clutch will engage and transmit drive

SHAPING MACHINES These machines are used for generating flat surfaces and make use of a single point tool attached to the end of a horizontally reciprocating ram. Cutting takes place on the forward stroke and the tool is lifted clear on the return stroke by means of a hinged *clapper box*. On some heavy duty machines the tool may be lifted clear using a power operated device. Figure 6.7(a) shows the basic design of a shaping machine.

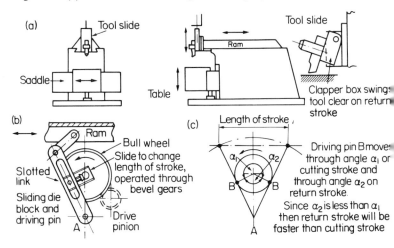

Figure 6.7 Shaping machine

The ram is operated by a quick return motion as shown in Figure 6.7(b). As the bull wheel rotates, the slotted link will oscillate about the pivot A. The length of stroke may be changed by adjusting the die block. This effectively increases or decreases the radius from the centre of the bull wheel to the centre of the driving pin. A feature of this mechanism is that the non-cutting return stroke is faster than the cutting stroke. This will be readily appreciated by considering Figure 6.7(c).

The work is clamped to the table and fed across the machine at right angles to the ram movement, which may be performed manually or under power. Power feeds use a ratchet and pawl mechanism operated by a drive from the bull wheel, so that the table is indexed laterally on the return stroke of the ram. The feed rate is determined by the setting

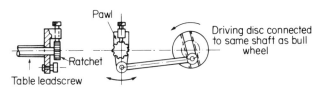

Figure 6.8 Shaping machine feed mechanism

radius on the driving disc as shown in Figure 6.8. Angular faces can be produced by inclining the tool slide and using a hand feed.

Shaping machines are not regarded as being suitable for quantity production, and hence are used mainly for prototype work, especially where large stock removal is required.

MILLING MACHINES Milling is a machining process, used to produce flat or profiled surfaces using a rotating multi-tooth cutter. For some applications a single point rotating cutter may be used. Both generating and forming operations may be carried out on these machines.

Milling machines are classified as being either *horizontal* or *vertical spindle* types.

Horizontal milling machines

The spindle axis of these machines lies in the horizontal plane. Fitted to the spindle is the *arbor*, on which the cutters are mounted. It is supported

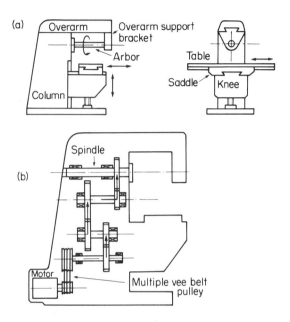

Figure 6.9 Horizontal milling machine, (a) general layout, (b) simplified view of transmission from motor to spindle

at the end by the arbor support bracket. Figure 6.9 shows the general arrangement for a horizontal milling machine together with the transmission from the input to the spindle.

TYPES OF CUTTER

The type of cutter used will depend on the surface/shape to be machined. The following are a representative range of milling cutters.

(1) *Slab* or *roller mill*, used for machining large flat surfaces.

(2) *Side and face cutters*, used for machining two faces at 90° to each other simultaneously. This is a useful cutter and finds many applications in milling operations.

(3) *Profile cutters* are designed and ground to a special shape for imparting a profile on to the work, e.g. convex or concave.

(4) *Angle cutters*, used for machining angles and vees, and may be either single or double angle. Typical angles for these cutters are 30°, 45° and 60°.

Figure 6.10 illustrates these cutters, together with some typical applications.

The range of work performed on horizontal machines may be greatly extended by having a table that swivels about the vertical axis. Such machines are known as *universal machines* and are often supplied with

D

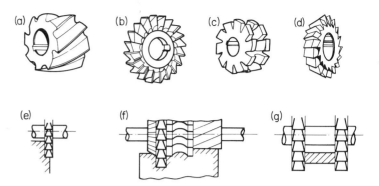

Figure 6.10 Horizontal milling cutters and applications, (a) slab or roller mill,
(b) side and face cutter, (c) profile cutter (convex), (d) angle cutter, (e) machining
a step using a side and face cutter, (f) gang milling, several cutters set up to
machine a complex profile, (g) straddle milling

attachments such as a vertical head and a slotting head. Universal milling
machines are frequently found in toolrooms due to the wide range of
operations that may be performed in them.

Bed type milling machines These machines, sometimes known as *manufacturing machines*, are of
more robust construction compared with horizontal machines and there-
fore more suitable for heavy duty production milling. The bed is of fixed
height and the cutter height is adjusted by moving the overarm and spindle. These machines are invariably used with milling fixtures, which locat

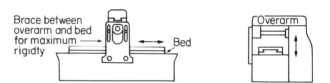

Figure 6.11 Bed type milling machine

the component precisely relative to the cutters. Figure 6.11 shows the
general arrangement of a bed type machine.

Vertical milling machines This is similar to the horizontal machine except that the spindle axis is
at 90° to the machine table. Figure 6.12 shows the general layout of
such a machine, together with the method of arranging the final drive
to rotate about the vertical axis.

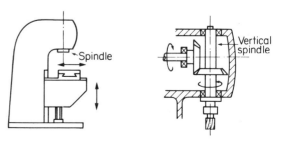

Figure 6.12 Vertical milling machine

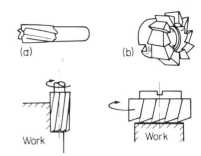

Figure 6.13 Vertical milling cutters and applications, (a) endmill, (b) shell mill, (c) use of endmill to machine a step, (d) use of shell mill to machine a flat surface

The types of cutters used on vertical machines are mainly *end mills* and *shell mills*, which are both designed to cut on the front and side. Shell mills are similar to end mills except that the former generally has a diameter larger than its length. These cutters are shown in Figure 6.13.

GRINDING MACHINES

In grinding operations material is removed using an abrasive wheel rotating at high speed. A grinding wheel may be regarded as a multi-tooth cutter, each abrasive particle acting as a cutting tooth. Grinding is essentially a finishing process carried out to provide close dimensional accuracy with improved surface finish on previously machined work. The amount of metal removed during the process is very small.

Grinding machines may be calssified into two main groups: *surface grinding machines* and *cylindrical grinding machines.*

Surface grinding machines

These machines are designed to generate flat surfaces and fall into two basic types, these being *horizontal* and *vertical spindle* machines.

In the horizontal machine shown in Figure 6.14(a), the grinding wheel is considerably narrower than its diameter and cuts on its periphery. The work reciprocates at right angles to the spindle axis and is also given a cross feed at the end of each pass. The wheelhead is attached to a vertical slide which permits the wheel height to be adjusted to alter the depth of cut. Coarse and fine adjustments are provided, the fine permitting cuts of the order of 0.002 mm to be applied. Figure 6.14(b) shows a section through the wheelhead of a horizontal surface grinding machine.

In the vertical spindle machine, the wheel is in the form of a ring and is used to cut on its front face. This arrangement results in a larger area of contact between the wheel and work, compared with the horizontal machine, and hence is likely to generate more heat. It is therefore important that a copious supply of coolant is provided to the wheel and work. These machines are of heavy construction and are therefore suited to production surface grinding. They are not, however, generally regarded as being suitable for grinding to such fine limits as can be achieved with horizontal surface grinders.

Work holding devices usually take the form of a magnetic chuck whose top surface is flat to receive the work. This of course is only suitable for ferrous metals; non-ferrous metals may be clamped directly to the machine table, or held in a suitable vice.

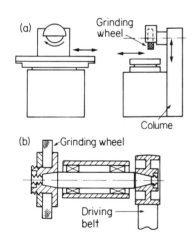

Figure 6.14 Horizontal surface grinding machine, (a) general layout, (b) simplified section through wheelhead

Cylindrical grinding machines

In cylindrical grinding the work rotates and reciprocates parallel to the axis of the wheel (*traverse grinding*). Alternatively, where the work is less than the width of the wheel, the work may be fed directly into the wheel (*plunge grinding*). Machines are classified as being either *plain* or *universal.* Plain machines are confined to the grinding of parallel diameters,

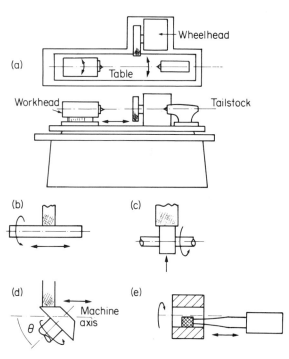

Figure 6.15 Universal grinding, (a) general layout, (b) traverse grinding, (c) plunge grinding, (d) grinding a short taper by turning the workhead through angle θ, (e) internal grinding

while universal machines have the added facility for internal grinding and also the grinding of tapers, both external and internal. The layout of a typical universal grinding machine is shown in Figure 6.15 together with the range of operations that may be performed.

DRILLING MACHINES

There are many variations of drilling machines in common use, but in general they may be classified into three basic types: *sensitive*, *pillar* and *radial*.

Sensitive drilling machine

This machine is designed for drilling holes of about 2–18 mm in diameter It is usually bench mounted and the drill feed applied by hand, so that the operator is able to sense the drill cutting. The drive from the motor to the spindle is in the form of a 'vee' belt and pulley. Change in speed is carried out by changing the position of the belt on a coned pulley. Figure 6.16 shows the general arrangement for a sensitive drilling machine

Pillar drilling machine

This machine is a larger version of the sensitive machine and is of more robust construction and hence more suited to the drilling of large diamete holes. The drive from the motor to the spindle is through a gearbox, whic provides about six to eight spindle speeds, ranging from about 50 to 1100 rev/min. Although the drill feed may be by hand, power feeds are also available. The pillar drilling machine is illustrated in Figure 6.17(a).

Radial arm drilling machine

Where work is either too heavy or bulky to be manipulated on a pillar drilling machine, a radial arm machine is used, as shown in Figure 6.17(b) In this machine the spindle is carried in a drill head which can be moved

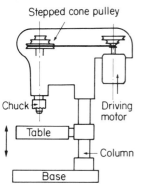

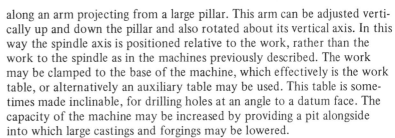

Figure 6.16 Sensitive drilling machine

along an arm projecting from a large pillar. This arm can be adjusted vertically up and down the pillar and also rotated about its vertical axis. In this way the spindle axis is positioned relative to the work, rather than the work to the spindle as in the machines previously described. The work may be clamped to the base of the machine, which effectively is the work table, or alternatively an auxiliary table may be used. This table is sometimes made inclinable, for drilling holes at an angle to a datum face. The capacity of the machine may be increased by providing a pit alongside into which large castings and forgings may be lowered.

The principle of the geared transmission for this machine is very similar to that of the pillar drilling machine and is illustrated in Figure 6.17(c).

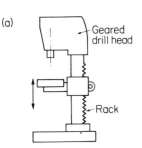

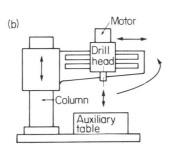

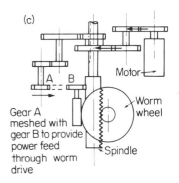

Figure 6.17 (a) Pillar drilling machine, (b) radial arm drilling machine, (c) transmission system

Drilling machine spindle

One design feature that is common to all drilling machines is the method by which rotation is combined with an axial *spindle* motion. A simplified spindle is shown in Figure 6.18. The axial motion is produced by the rack and pinion mechanism, the rack being cut into the sleeve carrying the spindle. An extension of the spindle passes through the centre of the pulley which permits the spindle to move axially along a keyway. On some machines a spline may be preferred, especially if high drilling torques are involved.

SLIDEWAY SYSTEMS

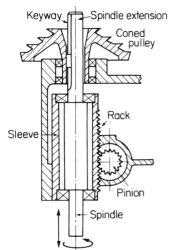

Figure 6.18 Drilling machine spindle

The function of a machine tool *slideway* is to locate and guide various moving members to enable the tool and work to be positioned relative to each other. This movement usually takes the form of a straight line motion.

All slideways should have provision for:

(1) Accurate alignment relative to other parts of the machine
(2) Freedom from unnecessary restraint
(3) Adjustment to take up wear
(4) Prevention of ingress of foreign matter between the sliding surfaces
(5) Lubrication.

Figure 6.19 shows some examples of slideways.

(a) *Vee and Flat* This is a common type of slideway often found on lathes, for guiding the tailstock and carriage along the bed. It is relatively easy to manufacture and is self adjusting for wear. It should be appreciated that if two vees were used, they would be difficult to manufacture in terms of getting an exact fit. In addition, it would also be difficult to provide adjustment if wear was uneven.

(b) *Dovetail* This slide is used where an upward lift is to be prevented.

Adjustment for wear is provided by means of a *gib strip.* Figure 6.19(b) shows two forms of this slide.

(c) *Use of rolling elements* The slides so far discussed will offer up considerable frictional force when moved, due to the large surface area of contact. This may be overcome by using balls and rollers between the slideway members, so that sliding friction is substituted for rolling friction

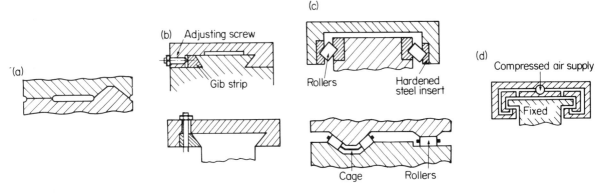

Figure 6.19 Slideway designs, (a) vee and flat, (b) dovetail, (c) use of rolling elements, (d) pneumatic slideway

This now provides the slide with light control together with sensitivity of movement, a feature sometimes required in certain precision grinding machines. Figure 6.19(c) illustrates two examples of this type of slideway

(d) *Pneumatic slideways* These slides were developed to further reduce friction between the sliding members. Figure 6.19(d) illustrates a pneumatic slide whereby the moving member floats on a thin film of compressed air. Alternatively pressurised oil may be used. Whichever medium is used, it is important that this film remains constant if the positional accuracy of the slide is not to be reduced.

SLIDE MOVING DEVICES

There are many slide moving devices used on machine tools, the following being the most common: *screw and nut, rack and pinion* and *hydraulic systems.*

Screw and nut

These devices are the most common and are widely used on lathes, milling and shaping machines. The device may be used in one of two ways: either by moving the screw or moving the nut. Figure 6.20 shows examples of these methods applied to the cross slide of a lathe and the knee elevating mechanism of a milling machine.

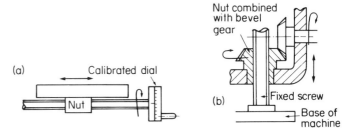

Figure 6.20 Screw and nut mechanisms, (a) cross slide mechanism for a lathe (moving screw, fixed nut), (b) elevating mechanism for a milling machine (moving nut, fixed screw)

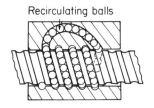

Recirculating balls

Figure 6.21 Recirculating nut

The screw may also be used as a method of controlling the movement of the slide by attaching a calibrated dial to the screw handwheel, as shown. If the pitch of the screw is known, then this dial may be divided into a suitable number of divisions to give fine positional control to the slide, e.g. – if the screw has a pitch of 6 mm and the dial is divided into 120 divisions then each division on the dial will represent $6/120 = 0.05$ mm.

To provide sufficient strength to machine tool leadscrews, threads having an acme or square form are used. The efficiency of these screws is low, of the order of 10–20%. Where higher efficiencies are required a recirculating ball screw and nut may be used, as shown in Figure 6.21. The efficiency of this mechanism is claimed to be greater than 90%.

Rack and pinion The purpose of a rack and pinion is to convert a rotary motion into a linear motion. The two most common examples of this are in the carriage operating mechanism of the lathe and the spindle movement on the drilling machine. The rack and pinion is also to be found on grinding machines in the cross feed mechanism, although for the table traverse a hydraulic system is to be preferred.

Hydraulic system Oil is virtually an incompressible fluid and provides a suitable medium for transmitting mechanical energy. The use of oil in a hydraulic system of a machine tool offers the following advantages:

(1) Smooth operation of reciprocating motions, e.g. as found in the shaping and grinding machine
(2) Wide range of speeds possible
(3) Quiet in operation
(4) Relief valves provide a safety factor
(5) Self lubricating.

The main disadvantages, however, are leaks and change in oil viscosity due to variations in temperature.

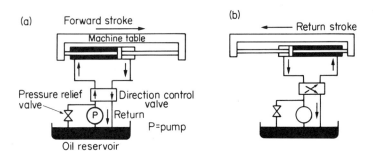

Figure 6.22 Simple hydraulic circuit applied to a surface grinding machine

Figure 6.22 shows a simple hydraulic circuit applied to a surface grinding machine.

SUMMARY Machine tools are used for shaping metal under power, either by using generating or forming principles. The centre lathe represents the basis for all machine tools from which others have developed.

Capstan and turret lathes are used for rapid manufacture of turned parts using pre-set tooling.

Shaping machines are primarily used for generating flat surfaces using a reciprocating single point tool.

Milling machines (horizontal and vertical) are used for producing flat or profiled surfaces using multi-tooth cutters.

Grinding machines (surface and cylindrical) are used for grinding previously machined work, using an abrasive wheel to give improved surface finish.

Drilling machines are of three types and include the sensitive, pillar and radial arm machines.

In any machine tool it is important that the various slides move accurately to each other and without any unnecessary restraint. The movement of these slides may be actuated by various methods; these include screw and nut, rack and pinion or the use of a hydraulic system.

QUESTIONS

(1) Explain using suitable sketches the difference between generating and forming.

(2) Describe with the aid of a suitable diagram the transmission from the motor to the tool and work for a centre lathe.

(3) Describe the function of capstan and turret lathes and explain the main difference between these two types of machines.

(4) Explain how the headstock on capstan and turret lathes differ from that of a centre lathe.

(5) Describe using a clear diagram the quick return mechanism as used in the shaping machine.

(6) Using a clear diagram show the transmission from the motor to the spindle for a horizontal milling machine.

(7) Make a clear diagram of the head of a vertical milling machine showing how the drive is transmitted to the vertical spindle.

(8) Using an outline diagram show clearly the difference in construction between a pillar drilling machine and a radial drilling machine.

(9) Show using a simplified diagram a typical layout of the motor to the spindle for a radial drilling machine.

(10) State three important features that a slideway system should possess.

(11) With the aid of a clear diagram show how the adjustment for wear may be provided for on a dovetail slide.

(12) Using a suitable sketch describe the advantages of a slideway using balls or rollers between the moving parts.

(13) With the aid of a suitable diagram show how a moving nut and leadscrew may be used in the knee elevating mechanism of a milling machine.

(14) Describe, using a suitable diagram, how the position of a machine tool table may be controlled using a headscrew and nut.

(15) With the aid of a diagram describe the principle of operation of a drilling machine spindle.

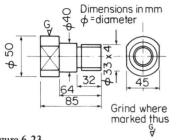

Dimensions in mm
φ = diameter

Grind where
marked thus
G
∇

Figure 6.23

(16) A milling machine table has a leadscrew of pitch 5 mm. The calibration dial is divided into 200 divisions. To what accuracy can this table be positioned?

(17) State three advantages that a hydraulic slide moving device has compared with a mechanical system.

(18) For the component shown in Figure 6.23 list in logical order the machining processes required for its manufacture.

7 Measurement

INTRODUCTION Measurement is of considerable importance in engineering and is required at all stages of manufacture.

Any system of measurement must be related to a known standard, otherwise the measurement has no meaning. Man has recognised the need for standards even as far back as 3000 BC when the earliest recorded length standard was the Egyptian cubit, which was the distance from the elbow to the tip of the outstretched middle finger. Throughout the centuries many other attempts were made to form length standards. These include the 'foot', which was defined as the length of the actual foot of the reigning monarch, and the English rod, defined as the total length of the left feet of the first sixteen men to come out of church on Sunday morning. These examples are of course only of historic interest and no doubt the reader may even find them amusing. Nevertheless, they were serious attempts of the period to provide a suitable standard of length.

It wasn't until 1855 that an accurate standard was made. This was known as the *imperial standard yard* and by Act of Parliament of that year was defined as the separation of two lines engraved on gold plugs, set into a bronze bar, under specified reference conditions. The French in 1875 established the international metric system and adopted the prototype *metre* as its standard. The metre was defined as the separation of two lines engraved on a special platinum/iridium bar under specified reference conditions. A copy of the standard metre together with the standard yard is kept in the United Kingdom by the National Physical Laboratory (NPL). Since the yard and the metre are material standards they are susceptible to the following disadvantages:

(1) It has been noted that over a period of years these standards have slightly changed in length.
(2) If these standards accidentally become damaged or destroyed then exact copies could not be made.

Due to these disadvantages, it was decided in 1960 to use light as a length standard, which can be reproduced without fear of variation. The metre is now defined as being equal to 1650763.73 wavelengths of orange radiation of krypton 86 under specified reference conditions.

Relationship between yard and metre Up until 1960 yard standards used by USA, Canada, Australia, New Zealand and South Africa differed slightly. These countries, together with the United Kingdom, decided to re-define the yard in terms of the metre. The standard yard is now known as the *international yard* and is defined as

1 international yard = 0.9144 metre

This also results in the inch being exactly equal to 25.4 mm.

Like the metre this means that the international yard will also be expressed in terms of the wavelengths of light. Thus

$$1 \text{ international yard} = 0.9144 \times 1650763.73 \text{ wavelengths of krypton 86}$$

At the time of writing the United Kingdom is undergoing a change from the Imperial system of measurement to the SI system (Systeme International d'Unites). When this change is complete the international yard will no longer be required and the metre will finally become the legal standard of length.

WORKING STANDARDS OF LENGTH

So that components may be manufactured relative to the legally defined standard, industry must use working standards. These consist of either *line* or *end* standards. Line standards take the form of accurately divided scales and are incorporated in measuring equipment and precision machine tools. End standards are the most common type of working standard and take the form of slip gauges (gauge blocks) and length bars.

SLIP GAUGES

These are rectangular blocks of high quality hardened and stabilised steel, whose length between parallel measuring faces has been made to a high degree of accuracy. In addition the measuring faces are ground and lapped to a very high degree of flatness and surface finish. A set of slip gauges has a range of sizes that enables dimensions to be produced by very small increments. The following table shows the gauges to be found in a 47 piece set.

Range (mm)	*Increments* (mm)	*Pieces*
1.005	0.005	1
2.01–2.09	0.01	9
2.10–2.90	0.1	9
1–24	1.0	24
40, 60, 80, 100	20	4
		47

Slip gauges are available in five grades of accuracy, classified as: calibration, 00, 0, I and II. The 0, I and II grades are intended for general use, grade 0 being most suitable for high class inspection work and the setting of measuring equipment. For workshop use, grades I and II are generally used. The calibration and 00 grades are reserved by inspection departments and standards rooms to enable other grades of slip gauges to be checked.

To obtain a required dimension, individual gauges may be *wrung* together, by bringing them into contact at right angles to each other and then sliding them through 90°, as shown in Figure 7.1. The term *wrung* or *wringing* is used to describe the effect when two slip gauges adhere together when their measuring surfaces come into close contact under pressure. If wringing is to take place easily, then the gauge faces must be absolutely clean and free from dust.

To obtain maximum accuracy from a slip gauge build up, the minimum number necessary should be used. As a general rule the smallest unit of the dimension should be considered first.

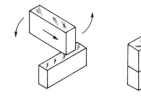

Figure 7.1 Wringing slip gauges together

Example Select gauges from a 47 piece set to build up 46.345 mm.

1st gauge	1.005
2nd gauge	2.04
3rd gauge	2.3
4th gauge	1
5th gauge	40
total	46.345 mm

Protector slips Grades 0, I and II sets are usually supplied with two extra gauges known as *protector gauges* having a length of 2 mm. One gauge is wrung on to each end of the build up, thereby ensuring that wear is confined only to two gauges. So that protectors are extremely resistant to wear they are often made from tungsten carbide and are usually identified by the letter P on one measuring face.

Slip gauge accessories These are available as sets comprising a base, special jaws and holders to extend the application of slip gauges. The use of these accessories enables a wide range of gauges and marking out tools to be assembled, two examples being shown in Figure 7.2.

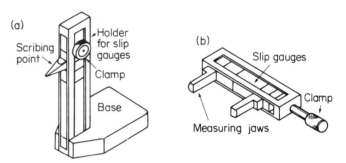

Figure 7.2 Slip gauge accessories, (a) height gauge, (b) gap gauge

Care of slip gauges The following should be observed to preserve the accuracy of slip gauges.

(1) It is important that the measuring faces are clean and undamaged before wringing gauges together. If difficulty is experienced then the measuring faces should be closely examined for small scratches.

(2) Measuring faces should not be fingered, so that the risk of tarnishing is minimised.

(3) Gauges should not be wrung together over an open box of gauges. If one is accidentally dropped then several gauges within the box could be damaged.

(4) Gauges should not be wrung together for longer than is necessary.

(5) Immediately after use, the gauges should be slid apart (not pulled), cleaned and the measuring faces smeared with a suitable protective grease.

LENGTH BARS Where end standards are required which are outside the range of slip gauges, *length bars* are used. These are of circular cross-section and vary in length from 25 mm to 1200 mm.

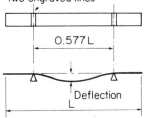

Figure 7.3 Length bar connection

Length bars are covered by BS 1790 which specifies four grades of accuracy: *workshop*, *inspection*, *calibration* and *reference*. From their names it is obvious where workshop and inspection grades of bar are used. The joining of these bars is achieved by using a loosely fitted screwed stud. This stud should be a loose fit so that adjacent bars come together easily and without unnecessary strain. Figure 7.3 shows the recommended method for joining workshop or inspection length bars.

Calibration and reference bars have completely plane faces and are only used for work requiring a high order of accuracy. Although the flatness and parallelism tolerance is the same for both bars, the only difference between them is that reference bars are given a closer tolerance on their length. Reference bars should only be used in standards rooms where the temperature is controlled at 20 °C ± ½ °C.

Supporting length bars

When length bars are used in the horizontal position they should be supported at two points and in such a way that no deflection occurs over their ends. In this way, the measuring faces will be parallel. The two points of support are known as the *Airy* points, named after Sir George Airy (1801–1892) who investigated the requirements for supporting length standards. These points are specified as being 0.577L, where L is the length of the bar, and are indicated on individual bars of 150 mm in length and over (Figure 7.4). Where length bars are screwed together as a combination, the separation of the Airy points must be determined by calculation.

SURFACE PLATES

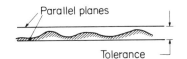

Figure 7.4 Supporting length bars

Surface plates are used as datums from which measurements are taken and they also provide a workshop standard for flatness. The material used for surface plates may be either of good quality cast iron or granite and should be of such construction that maximum rigidity is obtained. In cast iron plates this is achieved by using deep ribbing on the underside. Granite plates are made in the form of a rectangular solid block. Surface plates are classified according to three grades of accuracy, these being grades AA, A and B. Grades AA and A plates are used for high class precision and inspection work. Cast iron plates have their working surfaces finished by grinding and lapping. A scraped plate is to be preferred since measuring equipment, such as height gauges, will not wring to the working surface. Grade B cast iron plates may also have a scraped surface or may be left finished machined by planing using a broad nose tool. These plates are often found in workshops for marking out purposes.

The accuracy of a surface plate is specified by the total permissible variation of the working surface contained between two parallel planes. This is illustrated in Figure 7.5.

The flatness tolerance on a 1000 mm × 630 mm plate according to grade is shown in the table.

Figure 7.5 Surface plate accuracy

Grade	Tolerance (mm)
AA	0.006
A	0.012
B	0.048

Checking a flat surface

It is sometimes necessary to check the flatness of work by comparing it with a surface plate. This is done by smearing the surface plate with a thin film of *engineers' blue* (sometimes known as *Prussian blue*) and then lightly rubbing the two surfaces together. The blue will then be transferred on to the high spots of the work, which are then removed by scraping. The process is repeated until there is an even distribution of blued spots over the whole surface. This is a specialised process and accurate results can only be obtained by experience.

Generation of a flat surface

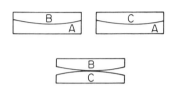

Figure 7.6 Generating a flat surface

Although we have just seen how a surface may be scraped flat with reference to a surface plate, the reader may wonder how the surface plate was scraped flat in the first place. During their manufacture cast iron plates are made in threes, the principle being that these plates will only mate together if all three surfaces are flat. Figure 7.6 shows this principle.

Plates A and B are scraped so that they bed together which is then followed by scraping A to C. If A is concave then B and C will become convex and will show up when these plates are blued together. B and C are then partially corrected by scraping and hence will both show the concavity of A. This sequence is repeated until the plates are flat within the specified tolerance.

STRAIGHTNESS

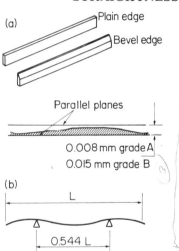

Figure 7.7 Straight-edges, (a) steel straight-edges, (b) straight-edge supported to give minimum deflection

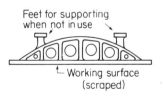

Figure 7.8 Cast iron straight-edge

A simple definition of *straightness* may be expressed as the shortest line connecting two points. Any departure from this line will therefore result in a straightness error. Standards for straightness come in the form of *straight-edges.* These consist of a length of steel or granite having a deep and narrow rectangular section. This type of section is essential if deflection in the plane of measurement is to be minimised. Small straight-edges up to about 200 mm in length have a bevelled edge and are sometimes known as *toolmakers' straight-edges.*

BS 5204 specifies two grades of accuracy, these being grade A and grade B. The tolerance for straightness is specified as the maximum permissible separation of two parallel planes between which the face lies. For example, a straightedge 1 m long has a tolerance of 0.008 mm and 0.015 mm for grades A and B respectively, as illustrated in Figure 7.7(a). To preserve this accuracy in use, long straight-edges should be used on edge and supported so as to give minimum deflection. This condition will only occur when the supports are separated at a distance 0.554L, where L is the length of the straight-edge, as shown in Figure 7.7(b).

Straight-edges may also be made from cast iron and may be likened to a long narrow surface plate. These are useful for checking the flatness and straightness of machine tool guideways. Figure 7.8 shows a typical cast iron straight-edge.

Steel bevelled straight-edges are often used by estimating the light gap between the straight-edge and the work. A light gap of 0.002 mm is easily detected while with experience and good eyesight a gap of 0.001 mm is detectable; in this case the light will be tinted blue since white light will not pass through a 0.001 mm gap.

SQUARENESS

Squareness is simply another term for 90°, which is a very important angle in engineering. For example, the accuracy of machine tools is partially dependent on spindles and various slides moving at 90° to each other.

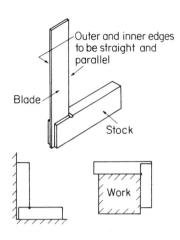

Figure 7.9 Engineers' square

The most common square found in the workshop is the *engineers' square*, shown in Figure 7.9. The inner and outer edges of the blade and stock must be parallel and straight, and firmly fixed at 90° to each other. BS 939 classifies engineers' squares into three grades of accuracy: AA (reference), A and B. Grade AA (reference) squares are the most accurate and should only be used for high class inspection work.

Cylindrical and *block squares* (see Figure 7.10) are also available which have the advantage of being more robust and hence are more likely to maintain their accuracy over a long period. Cylindrical squares are available in one grade only, this being AA (reference), while block squares are available as either grade A or grade B. Both cylindrical and block squares are ideal for checking the accuracy of engineers' squares.

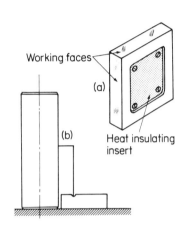

Figure 7.10 (a) Block square, (b) checking on engineers' square against a cylindrical square

ANGLE MEASUREMENT

The units of angular measurement are the degree, minute and second and are related as follows:

$$1 \text{ degree} = 60 \text{ minutes}$$
$$1 \text{ minute} = 60 \text{ seconds}$$

There are many methods of measuring angle, the choice of which will depend upon the accuracy required from the measured result. The *vernier bevel protractor* is perhaps one of the most common workshop instruments used for angle measurement, giving a direct reading accuracy of 5 minutes. Many manufactured components, however, require much greater accuracy, for which some of the following methods may be suitable.

Angle gauges These consist of precision gauge blocks having two inclined faces forming a known angle and which has been produced to a high degree of accuracy. The surface finish and flatness of these faces are similar to those of slip gauges, thus permitting angle gauges to be wrung together to build up a required angle. A complete set of angle gauges consists of thirteen pieces which will enable any angle to be built up between 0 and 360° in steps of 3 seconds.

A complete set of angle gauges consists of the following pieces:

Degrees	1	3	9	27	41
Minutes	1	3	9	27	
Seconds	3	9	27		

plus 90° in the form of a square block.

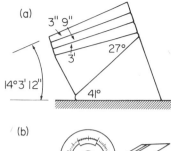

(a)

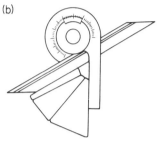

(b)

Figure 7.11 (a) Angle gauges, (b) checking a vernier protractor using angle gauges

To build up a required angle, individual gauges are either added or subtracted. Gauges which are subtracted are simply wrung together the reverse way round. The following example will demonstrate this principle.

Example Show what gauges are required to build up an angle of 14° 3′ 12″.

By treating the degrees, minutes and seconds separately then

$$41° - 27° = 14°$$
$$9″ + 3″ = 12″$$

Note that 3′ is available as a separate piece.
These gauges will then be wrung together as shown in Figure 7.11(a).

Angle gauges are used as angle standards and provide a convenient method of checking the accuracy of measuring equipment such as a vernier protractor as shown in Figure 7.11(b).

The handling and care of angle gauges should be the same as that exercised for slip gauges.

Sine bar

The *sine bar* basically consists of a ground bar into which is fitted two rollers at a known centre distance. When used in conjunction with slip gauges the sine bar may be used to set up or measure an angle, as shown in Figure 7.12.

It will be apparent from Figure 7.12 that the geometry of the sine bar is based on a right angle triangle. If h is the height of the slip gauge pile and l is the centre distance between the rollers then

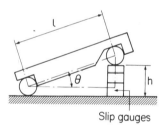

Figure 7.12 Sine bar

$$\sin \theta = \frac{h}{l}$$

If a sine bar is to be accurate then the following conditions must exist.

(1)   The axes of the rollers must be parallel to each other and the centre distance, l, must be precisely known. The size of a sine bar is specified by this distance.

(2)   The top surface of the bar must be flat and parallel to a plane connecting the axes of the rollers.

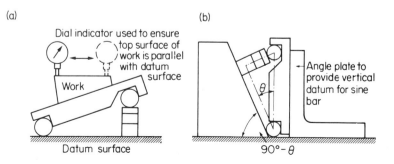

Figure 7.13 Measurement of angle greater than 45°, using a sine bar

(3) The rollers must be of identical diameters and round to within a close tolerance.

The most common method of measuring an angle using a sine bar is to raise the front end with slip gauges until the work surface is parallel to the datum surface as shown in Figure 7.13(a). Since the angle of the component will be near to the nominal size, the theoretical slip gauge pile is usually calculated first, which can then be adjusted until the work surface is parallel with the datum surface.

Errors in sine bars are greatly magnified if they are used to measure angles greater than 45°. For this reason it is usual to measure the complementary angle, which may be subtracted from 90° to obtain the required angle. This is illustrated in Figure 7.13(b).

Sine centres

Due to the difficulty of mounting conical work easily on a conventional sine bar, *sine centres* are used. This equipment consists of a self-contained sine bar, hinged at one roller and mounted on its own datum surface. The top surface of the bar is provided with a pair of centres for holding the work, as shown in Figure 7.14. Due to the work being held axially between centres the angle of inclination will be half the included angle of the work.

The use of sine centres provides a convenient method of measuring the angle of a taper plug gauge.

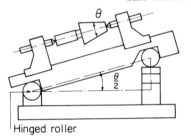

Hinged roller

Figure 7.14 Sine centres

Use of balls and rollers

Angular work such as taper gauges may be accurately measured by taking micrometer readings over balls or rollers, sometimes in conjunction with slip gauges. Balls or rollers are used to provide point or line contact at the measuring faces, which cannot otherwise be obtained when using a micrometer directly on an angular component. By applying simple right angle trigonometry the required angle may be calculated using various measured dimensions.

MEASUREMENT OF A TAPER PLUG GAUGE
A measurement l_1 is first made over two precision rollers (Figure 7.15). These rollers are then positioned on top of slip gauges of height h and then another measurement made (l_2).

Considering the triangle ABC,

$$\tan \theta = \frac{AB}{BC}$$

where θ = half the included angle of the taper.

$$AB = \frac{l_2 - l_1}{2} \qquad BC = h$$

Therefore

$$\tan \theta = \frac{l_2 - l_1}{2h}$$

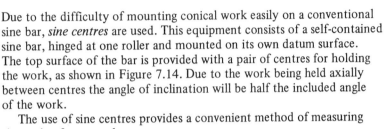

Figure 7.15 Measurement of an external taper

When θ has been found from the above expression it is simply doubled to give the required measured angle.

MEASUREMENT OF AN INTERNAL TAPER
The angle of an internal taper may be determined by taking readings with

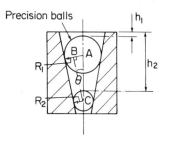

Figure 7.16 Measurement of an internal taper

a depth micrometer over the two precision balls. Figure 7.16 shows the relevant geometry.

$$\sin \theta = \frac{AB}{AC}$$

where θ = half the included angle of the taper.

$$AB = R_1 - R_2 \qquad AC = (h_2 + R_2) - (h_1 + R_1)$$

Therefore

$$\sin \theta = \frac{R_1 - R_2}{(h_2 + R_2) - (h_1 + R_1)}$$

ROUNDNESS

Many circular components that are produced by machine tools are not necessarily round, even though they may appear to be when measured. One such error in roundness is known as *lobing*. A figure that is said to be lobed is one which has an odd number of circular sides, but when measured between two parallel faces (e.g. a micrometer) appears to have

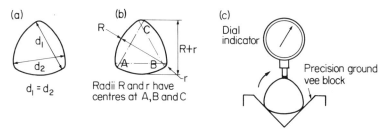

Figure 7.17 Lobing

a constant diameter, as shown in Figure 7.17(a). The most common lobed diameter found in engineering has three sides based on an equilateral triangle. By studying Figure 7.17(b) the reader will appreciate this constant diameter effect.

A diameter suspected of being lobed may be checked by rotation in a *vee block* under a dial indicator as shown in Figure 7.17(c). Certain angles of vee may not, however, disclose lobing, in which case two vee blocks having different angles should be used.

COMPARATIVE MEASUREMENT

Equipment such as rules, vernier calipers and micrometers are all *direct* measuring instruments since the reading obtained by them will be the

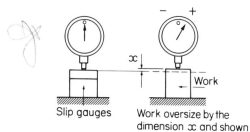

Figure 7.18 Comparative measurement

size of the measured part. The main disadvantage of these instruments is that they all rely on a zero datum, which if suspect will lead to an error in the measured result.

Comparative measurement minimises these errors and is a technique where an instrument is first set to a known standard, to which the work is then compared. The principle is shown in Figure 7.18 using a *dial indicator* as a simple comparator.

The standard chosen should be equal to the nominal dimension of the component, e.g. for a component specified as 15 ± 0.1 mm, use a 15 mm standard. This will ensure that the plunger of the dial indicator moves only by a small amount, which will help in reducing instrument errors.

Bench micrometer This instrument can only be used as a comparator and has the following advantages over a hand micrometer.

(1)   Large diameter thimble permits greater number of divisions around circumference, thus promoting better accuracy.

(2)   The fixed anvil is replaced by a fiducial indicator to ensure constant measuring pressure. This device is more reliable than the ratchet.

(3)   Micrometer screw errors will have minimal effect, since the screw is used over a very small range during measurement.

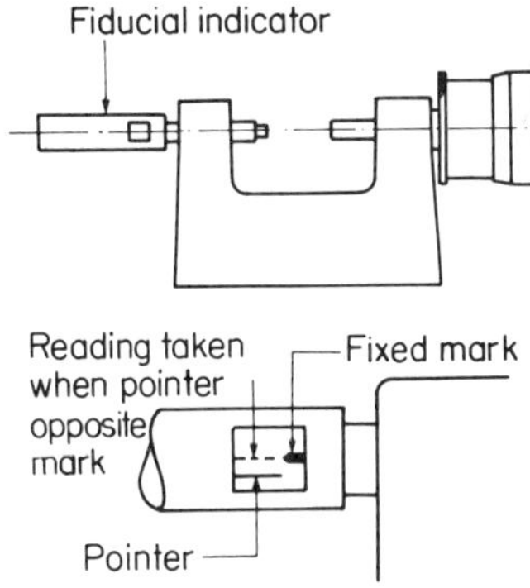

To measure the size of a component with the bench micrometer the following procedure is adopted. A suitable standard (S) is chosen, which will be near to the nominal size of the component. A reading (R_1) is now taken over the standard followed by a reading (R_2) over the component. The difference in the readings R_1 and R_2 will be the size difference between the standard and the component. The actual size (x) of the component will then be

$$x = S + (R_2 - R_1)$$

This expression assumes that the standard is smaller than the component. If larger then the expression becomes

$$x = S - (R_1 - R_2)$$

Although the dial indicator and bench micrometer are used for comparative measurement, there are specific instruments called *comparators*. These instruments are extremely sensitive, and embody different scientific principles to enable measured errors to be greatly magnified. Comparators will not be discussed in depth in this chapter, since they are dealt with more fully in level III work.

Figure 7.19 Bench micrometer, (a) overall view, (b) enlarged view of fiducial indicator

FURTHER APPLICATIONS OF THE DIAL INDICATOR

In addition to being used as a simple comparator to measure linear errors, the dial indicator is a useful piece of equipment for detecting geometrical errors. Figure 7.20 shows some typical examples.

Alignment of dial indicators

When in use it is important that the plunger axis of a dial indicator lies at 90° to the work surface. If this axis lies at some other angle then the plunger movement will not record the true error in the work. A reading error will now be produced, known as a *cosine error*, as shown in Figure 7.21.

Care should also be observed when using lever type dial indicators. The stylus of this instrument should be so arranged that the true line of

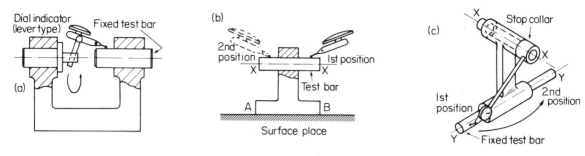

Figure 7.20 Geometry testing using a dial indicator, (a) checking the axial alignment of two bores, (b) checking the parallelism of the hole, axis x-x, with the base AB, (c) checking the squareness of the hole, axis x-x, to the hole, axis y-y

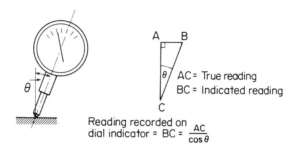

Figure 7.21 Cosine errors

measurement is tangential to the arc produced by the centre of the ball ended stylus. Most modern lever type indicators have a pear shaped stylus, which tends to minimise cosine errors.

CALIBRATION OF MEASURING EQUIPMENT

No matter how accurate a piece of measuring equipment is said to be, it will always have certain inherent errors, which may be caused by wear or possibly misuse. It follows then that some form of periodic check is required to determine the extent of these errors, which can then be allowed for in the final measurement. An instrument check of this nature is known as *calibration*. The principle of calibration is to compare the reading obtained by an instrument with that of a known standard.

Calibration of a 0–25 mm micrometer

The accuracy of a micrometer is largely dependent upon its screw thread and may be calibrated by taking readings over a series of slip gauges. A micrometer screw usually possesses two types of errors: *progressive* and *periodic*. Periodic errors are those that occur at regular intervals and may also be attributable to an eccentrically mounted thimble. Since progressive and periodic errors may exist together a suitable range of slip gauges must be chosen so that both types of error can be detected. A suitable range is as follows:

2.5 5.1 7.7 10.3 12.9 15.0 17.6 20.2 22.8 and 25.0 mm.

This range will not only check the micrometer on complete turns of the thimble, but also at intermediate positions. Readings are now taken with the measuring faces together, and then over the slip gauge series shown. Any differences between the reading and slip gauges will therefore be the

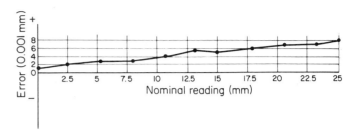

Figure 7.22 Typical calibration chart for a 0–25 mm micrometer

error in that particular reading. These errors are then plotted, to form a calibration chart as shown in Figure 7.22.

This chart can now be used to obtain an accurate measurement, e.g. for a reading of 10 mm the true dimension between the measuring faces will be

$$10.00 - 0.004 \text{ mm}$$
$$= 9.996 \text{ mm}$$

Calibration of a dial indicator   This is achieved in a similar fashion as with the micrometer. The dial indicator is fixed rigidly to a suitable stand and slip gauges built up and the corresponding reading noted. In the case of plunger dial indicators, it is usual to take readings at a tenth of a turn intervals during each revolution of the pointer. For example, an instrument having 100 divisions of 0.01 mm would be calibrated at 0.1 mm intervals throughout the range of the plunger movement. A calibration chart can now be constructed by plotting reading error against turns of the pointer. During this test the plunger should be carefully lifted and gently lowered on to the slip gauges, rather than slide the gauges underneath, which could otherwise strain the plunger unnecessarily.

Calibration charts for dial indicators often show periodic errors, in that the characteristic of the chart repeats itself for each turn of the pointer. These errors are often attributable to centring errors in the teeth of the pinion to which the pointer is attached, or to the axis of rotation of the pointer being not truly central with the dial.

SUMMARY   Working standards of length come in the form of slip gauges and length bars. Length bars over 150 mm should be supported at the Airy points. Surface plates provide a datum surface from which measurements may be made and also serve as a standard for flatness. Surface plates may be made of granite or cast iron.

Straight-edges act as a standard for straightness and are usually made of steel having a deep rectangular section.

Standards for squareness (90°) come in the form of engineers', block and cylindrical squares.

Angles may be measured using gauges and sine bars, or calculated from measurements taken over precision balls or rollers.

Some diameters (external) may be lobed, which is a constant diameter figure that is not round.

Comparative measurement is a technique for comparing the size of a component with a standard of known size.

Dial indicators are useful measuring tools for measuring geometrical errors such as alignment, parallelism and squareness.

All types of measuring equipment have inherent errors and should be calibrated to determine the extent of these errors.

QUESTIONS

(1) Describe how slip gauges should be wrung together, stating any necessary precautions that should be observed.

(2) Explain the function of protector slips.

(3) (a) Explain the importance of correctly supporting length bars in the horizontal position.

(b) Determine the separation of the points of support for a 750 mm length bar.

(4) Explain, using a sketch, how the flatness of a surface plate is specified.

(5) A machined surface is to be scraped flat. Explain how this may be carried out, using a surface plate as a reference surface.

(6) Using clear diagrams, describe suitable pieces of equipment for confirming (a) squareness, and (b) straightness.

(7) Explain what is meant by a diameter that is said to be lobed. Show using sketches how a lobed diameter may be detected.

(8) Explain clearly the difference between direct and comparative measurement.

(9) Using sketches, show how a dial indicator may be used as a comparator.

(10) A 12 mm diameter (nominal) plug gauge was measured using a bench micrometer and the following results were recorded.

Calibrated diameter of standard	10.00 mm
Reading over setting standard	6.61 mm
Reading over plug gauge	8.62 mm

What was the actual diameter of this plug gauge?

(11) State two advantages of using a bench micrometer compared with a hand micrometer.

(12) Using sketches show how the angle gauges may be built up to produce an angle of $36° 24' 6''$.

(13) A 200 mm sine bar is to be set up to an angle of $25° 30'$. Determine the slip gauge dimension required.

(14) Precision balls of diameter 20 mm and 15 mm were used to measure a taper ring gauge. The heights from the top surface of the gauge to the top of the large and small balls were 3 mm and 40 mm respectively. Calculate the included angle of taper for this gauge.

(15) Describe using a sketch how sine centres may be used to measure the included angle for a taper plug gauge.

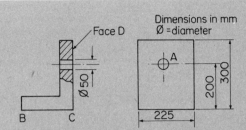

Figure 7.23

(16) With reference to the component shown in Figure 7.23 show with the aid of a diagram how a dial indicator may be used to inspect the following:

    (a) Parallelism of hole A with the base BC
    (b) The squareness of the hole A with the face D.

(17) Explain the meaning of the term calibration when applied to measuring equipment.

(18) Describe how a micrometer or vernier caliper may be calibrate and show how the results may be represented on a calibration chart.